Subfossil Chironomidae of the Western and E

Subfossil Chironomidae of the Western and Eastern Carpathians

An Identification Guide to Larval Remains from Lake Sediments

By

Peter Bitušík
Ladislav Hamerlík
Tímea Chamutiová

BRILL

LEIDEN | BOSTON

Cover illustration: Mysterious morning by Lake Nižné Temnosmrečinské pleso in the High Tatra Mountains, Slovakia. With an area of 12 ha, the lake is almost 40 m deep and is situated in the subalpine zone (1674 m a.s.l).

The Library of Congress Cataloging-in-Publication Data is available online at https://catalog.loc.gov
LC record available at https://lccn.loc.gov/2024039125

Typeface for the Latin, Greek, and Cyrillic scripts: "Brill". See and download: brill.com/brill-typeface.

ISBN 978-90-04-69466-8 (hardback)
ISBN 978-90-04-69478-1 (e-book)
DOI 10.1163/9789004694781

This book is printed on acid-free paper and produced in a sustainable manner.

Dedicated to our children, Viktória, Martin, and Hanna, and to past, present, and future students

Contents

Foreword

When I was a student and I first heard the word "paleolimnology", the few publications in this field were mainly associated with sediment core analyses using pollen, diatom and/or geochemical analyses to reconstruct long-term trends in lake systems. Occasionally some papers would expand into chitinous zoological indicators, such as cladoceran exoskeletal remains and, even more rarely, the head capsules of chironomid larvae. This, however, all started changing in the late-1980s with many new applications for paleolimnology requiring the use of a wide spectrum of indicators, not least of which were with chironomids.

My personal connection with chironomid subfossils mainly began with the arrival of post-doctoral fellow Ian Walker to my newly established lab at Queen's University. Ian had been watching the development of a wide spectrum of calibration and numerical approaches that we were using with diatoms and chrysophytes to develop pH inference models for our acid rain research. Ian was already providing convincing (albeit qualitative) evidence that chironomids were sensitive to climate indicators, and specifically temperature. Applying the multivariate approaches used in our acid rain work, and a surface sediment calibration set of lakes from sub-Arctic Labrador (purposely chosen as they were almost all polymictic and so we could remove the effects of changes in deepwater oxygen), we developed the first temperature transfer functions using chironomid assemblages (Walker et al. 1991). Initially there was considerable opposition to these ideas, but as more and more researchers developed their own calibration sets, the idea that chironomids were important temperature proxies caught on. In fact, I would at times joke that people believed the Walker et al. paper too enthusiastically – we always recognized that there were significant caveats with this approach, not least of which was the importance of oxygen concentrations. The arrival of Roberto Quinlan in my lab further addressed the oxygen issue with calibration sets and applications for reconstructing deepwater oxygen levels (e.g., Quinlan and Smol 2001). A lot has happened in the intervening decades and chironomid subfossils are now used in a wide variety of applications.

However, in order to use chironomid assemblages in paleolimnological research, the critical first step is to correctly identify these indicators to the lowest taxonomic level possible. Whilst many bookshelves can be filled with volumes on, for example, diatoms, with few exceptions (e.g., Brooks et al. 2007; Andersen et al. 2013a,b) relatively little taxonomic information is available for chironomid subfossils. This new volume, summarizing subfossil Chironomidae from 63 Western and Eastern Carpathian lakes, fills an important void for taxa from mountainous regions of eastern Europe. Of course, subfossil chironomid taxonomy guides, such as this one, have great utility far beyond the geographical region of the lakes from which the remains were identified, as many subfossil chironomid taxa have wide continental-scale geographic distributions, at least at the genus level. The keys and many high-quality photographs characterizing this book are critical for accurate identifications. Importantly, the authors also provide images of head capsules from both living and subfossil material – demonstrating what taxonomic structures may be lost in the paleolimnological record.

This is an important book that should be placed next to the microscope of every scientist working on subfossil chironomids. As many new paleolimnological applications become developed, and especially for chironomid-based studies, research using these indicators will only continue to increase.

John P. Smol
Paleoecological Environmental Assessment and Research Lab (PEARL),
Dept. Biology, Queen's University, Kingston, Ontario, Canada
smol@queensu.ca

Preface

This book is the result of our decades-long palaeolimnological research in the Carpathians. It reviews chironomid larval remains found in the sediments of 72 lakes and presents identification keys to chironomid morphotypes, including their diagnosis, distribution, ecology, and palaeolimnological interpretation, along with rich photographic material. The illustrations and photographs are important parts of the book; we included several high-quality pictures of each type, showing the entire head capsule and main diagnostic features, usually the mentum and mandibles, but also other structures that can be useful for distinguishing individual genera and morphotypes if they are included in the larval remains. To demonstrate how many subtle morphological structures are missing in the subfossil remains compared to living larvae, illustrations show the diagnostic characters of both recent (left) and subfossil (right) head capsules. Chapters dealing with individual subfamilies are marked with different colors, making navigation in the book easier.

The identification keys presented in the book distinguish subfamilies, genera, and morphotypes recorded in our material from lakes in the Carpathians but will not work for taxa not included in it (e.g., the tribe Pseudochironomini). There are separate identification keys for subfamilies/tribes and for genera of each subfamily. One of the most important characters used in the general keys is the number of median teeth on the mentum; however, sometimes the exact number of median teeth is not obvious or can be confusing. Thus, genera with an unclear number of median mental teeth are keyed twice. In Orthocladiinae, these genera are *Brillia* (with both 2 and 3 median teeth) and *Diplocladius* (with 2 and 4 median teeth), and in Chironominae, *Microtendipes pedellus*-type (with both 2 and 3 median teeth).

Individual genera are presented using the following structure:

Diagnosis. This section presents the basic morphological features of genera. It includes structures that are frequently not present in subfossils, such as antennae, premandibles, or labral lamellae, especially when they are distinct for the genus. These characters may be useful when identifying living larvae or well-preserved subfossil remains. The diagnosis follows Brooks et al. (2007) and Andersen et al. (2013).

Remarks. This section presents similar genera and morphotypes, highlighting distinct diagnostic characters to distinguish one from another.

Morphotypes. If there are multiple morphotypes within a genus, we provide an identification key for their distinction. For new morphotypes, detailed descriptions are included. The distinction of morphotypes follows Brooks et al. (2007) unless stated otherwise.

Distribution. The distribution is entirely based on our data and shows the distribution of genera and morphotypes within the studied lake districts (orographic units, mountains). Countries where the morphotypes were found are marked as SK (Slovakia), PL (Poland), UA (Ukraine), and RO (Romania).

Ecology and Palaeolimnological Interpretation. When discussing ecological requirements, we summarize the basic knowledge about the genus and morphotypes using literature. The palaeolimnological interpretation of morphotypes is based on both published data from other researchers and our unpublished data and experiences.

In addition to chironomid morphotypes, the final chapter presents the most frequent non-chironomid insect remains found in lake sediments, summarizing the basic knowledge on the ecology and palaeolimnological application of different insect groups. This chapter also includes pupal remains of chironomids, specifically thoracic horns of Tanypodinae, which are frequently recorded in lake sediments.

At the end of the book, we present a list of chironomid genera that were not present in our subfossil material but whose presence in lake sediment in the Carpathians is very likely (Appendix 2).

We hope that you find the book useful. If you discover any mistakes or confusing information, please don't hesitate to contact us.

The authors

Acknowledgements

This book would not exist without the help of many people. We are grateful to all our colleagues and friend who helped us taking or processing lake-sediment samples, namely Milan Novikmec, Marek Svitok, Danka Dobríková, Veronika Slobodníková, Marcela Přidalová-Sedlačková, Lucia Nižníková, and Marcela Adamcová. Guillermo de Mendoza and Pete Cranston are acknowledged for their notes on *Ablabesmyia*; Arnold Móra for his valuable comments to the earlier version of the Tatra Mts. part of the manuscript. We are grateful to Dan Vondrák, Aneta Formáčková and Jesper Petersen for providing pictures of non-chironomid remains, Katarína Trnková of her photograph of lake Morské oko, and Nera Vuić for allowing us to use her drawing of a chironomid dorsal head. Help from Jozef Oboňa (Psychodidae), Matúš Kúdela (Simuliidae) and Oliver Heiri with the identification of non-chironomid dipteran remains is highly acknowledged. The temperature optima used in the book were calculated using the Tatra Mts. training-set by Marek Svitok. Proofreading of Martina Jambrović improved the clarity of the text significantly, *hvala*. Many thanks to Peter Turis for valuable comments to the structure of the Carpathians and for creating the map of the mountain range. The day-by-day support and patience of our partners and spouses during the never-ending writing of the book was substantial; thank you all.

This book is a result of a long-term research and received funding from various projects, such as APVV-20-0358, APVV-15-0292, VEGA: 1/0341/18, VEGA 1/0400/21, Austrian-Czech Lead Agency project – Austrian Science Fund FWF No. 19770 / Czech Science Foundation No. 20-00892L, Polish National Centre of Sciences Grant No. N N306 077436.

Figures

CHAPTER 1

Introduction

1.1 The Carpathians

The Carpathians (or Carpathian Mountains) form a huge arc of high mountains, plains, hills and valleys across Central and East Central Europe. Covering approximately 200,000 km^2 and stretching about 1,500 km, they represent one of the largest and most rugged mountain systems in Europe (Kadlečík et al. 2016).

FIGURE 1.1 Map of the Carpathians showing its main subunits
© PETER TURIS

 | DOI:10.1163/9789004694781_002

Apart from these exceptionalities, the Carpathians are an important biodiversity hot spot (see Mráz & Ronikier 2016 and references therein). In the European context, they have relatively well-preserved nature, hosting the largest natural forests and many relatively undisturbed non-forest ecosystems. Many semi-natural, traditional landscape types, habitats, plant and animal species represent distinctive or unique elements occurring mainly or exclusively in the Carpathian region. The mountains also provide habitat for the largest and most vital populations of large carnivores in Europe, along with top bird predators (Vološčuk et al. 1999, Oszlányi et al. 2004, Kadlečík et al. 2016).

1.1.1 *Physiography*

Even though the overall borders of the Carpathians are inconsistent and often being discussed, conventionally, the Danube River borders the Carpathians with the Danube Gap near Bratislava, Slovakia, being the northwest end and the Iron Gate near Orşova, Romania, being the southeast end of the mountain arc. However, the Carpathians extend across the Danube at both ends and eight countries: Austria, Czech Republic, Poland, Slovakia, Ukraine, Hungary, Romania and Serbia. More than 50% of its area is situated in Romania, followed by Slovakia (17%), Poland and Ukraine (10% each), while 5% and less is in the remaining countries (Kondracki 2022).

The Carpathians look like a "mountain island" surrounded by the Pannonian Basin and the East-European Platform. The valley of the rivers March and Oder separates the Carpathians from the Bohemian Massif. The Leitha Gate in Austria separates the Carpathians from the Alps, and the structural depression of the Timok River in Serbia separates them from the Balkan Mountains (Kondracki 2022). This delimitation of the Carpathians seems to be generally accepted, although some authors (e.g., Cankov 1974) believe that from the geological aspect, highlands in north-western Bulgaria could also be assigned to the Carpathians.

The Carpathians are traditionally divided into subunits based on geological, geographical and biogeographical aspects. Older studies have recognized two or three subunits: the Western and Eastern (South-eastern) Carpathians or the Western, Eastern and Southern Carpathians. Kliment et al. (2016) analysing an amount of relevant literature have distinguished five equivalent subunits within the Carpathians: the Western Carpathians (Austria, Czech Republic, Hungary, Slovakia, Poland), the Eastern Carpathians (Poland, Slovakia, Ukraine, Romania), the Southern Carpathians (Romania, Serbia), the Apuseni Carpathians (Romania) and the Transylvanian Basin (Romania) (Fig. 1.1).

It should be noted that the internal zonation of the Carpathians is not definitive and is constantly being modified. The situation is complicated also since the zonation of the eastern and southern parts is influenced by local traditions. For example, the section of the Carpathians within the borders of Ukraine is commonly known as the Ukrainian Carpathians, and Romanians sometimes denote, in local use, as the "Eastern Carpathians" only the Romanian part of the Eastern Carpathians, while they use their own classification scheme for it, and this scheme is also often used in scientific literature.

1.1.2 *Origin, Geomorphology, and Geology*

The Carpathians, as a part of the Alpine-Himalayan orogenic system, evolved during the Alpine Orogeny between 65 and 2.5 million years ago, although it is still active. This young orogenic system resulting from the Eurasian plate collision with the continental fragments of Gondwana after closing of the Tethys Ocean, is composed of many mountain systems, including the North African Atlas, through Pyrenees, Apennines, Alps, Carpathians, as well as mountain ranges in the Balkans and Caucasus to the Himalayas (Marko et al. 2020).

Even though the Carpathians are an eastern continuation of the Alps, they differ considerably from them. Above all, the average elevation the Carpathians is much lower than that of the Alps. Multiple peaks in the northwest (Slovakia) and south (Romania) parts reach a height over 2,500 meters, with peak Gerlachovský štít (2655 m), Slovakia, being the highest. Consequently, glaciers and extensive permanent snow fields that are common in the Alps, are absent in the Carpathians. The Carpathians are also more structured orographically compared to the Alps. They are divided into several mountain blocks separated by tectonic depressions and transversal valleys.

Geological structural elements are also different. While the sandstone and shale belt (flysch) is developed as a narrow belt on the northern edge of the Alps, in the Carpathians it is much wider and forms the main component of their outer zone. On the contrary, whereas the limestone rocks form a wide band in the Alps, they have secondary importance in the Carpathians. Similarly, the crystalline and metamorphic rocks are developed as robust chains in the central part of the Alps, whereas in the Carpathians they are present as isolated blocks surrounded by structural depressions. In addition to these characteristics, the Carpathians contain mountain groups of volcanic origin. The volcanic chain, stretching from Slovakia through Hungary and Ukraine to Romania with some huge extinct volcanoes, is the longest in Europe (Mahel' 1986, Kondracki 1989).

The differences are also visible in the relief of both mountain systems. During the last Ice Age, the Alps were largely glaciated (about 79% of the total area, ~150 000 km^2; Král 2001) and this gave them specific glacial features. In the Carpathians, glaciation affected only the highest parts (~0.5% of the total area, ~1100 km^2), so the recent relief forms are mostly results of flowing water action.

1.1.3 *Climate*

The geographical position of the Carpathians between western and eastern Europe, determines its climate. In winter, weather is governed by the inflow of polar-continental air masses arriving from the east and northeast, while during the rest of the year oceanic air masses flowing from the west play a dominant role.

Generally, the Carpathian climate possess certain features of continentality. The large distance from the Atlantic Ocean and the influence of the Alps and the Bohemian Massif cause a decrease in precipitation in the Carpathians, although the mountain range is much wetter than the surrounding plains (Kondracki 2022). The high mountain zone has a cold and moist climate with yearly average temperatures between +2 °C and −2 °C, with the annual precipitation from 1800 to 2000 mm in the north-western part of the Carpathians (in the Tatra Mountains even 2400 mm). In the Eastern and Southern Carpathians, precipitation ranges between 1400 and 1600 mm; however, in the Transylvanian Basin, the total annual precipitation is less than 600 mm per year.

Snow cover is present in the high mountains for 150 to 220 days of the year; however, permanent snow fields may be found all year round in the Tatra Mountains and the Rodna Mountains (Kadlečík et al. 2016).

1.1.4 *Hydrology*

Most of the Carpathians (~90%) belongs to the Black Sea-drainage area as most of the water runoff drains into the Danube, and in the east, in Ukraine, to the Prut and the Dniester rivers. Only the northern slopes of the Carpathians, situated mostly in Poland and partly in Slovakia and Czech Republic, are linked with the Baltic Sea through the catchment of the Vistula and the Oder.

The largest rivers that originate in the Carpathians include Vistula, Dniester and some tributaries of the Danube, such as Váh, Tisa, Olt, Mures, Siret and Prut. The Carpathian rivers are characterized by a rain-snow regime, so the highest discharges occur in the spring months, generally in March–April, but the high mountain streams attain the peak discharge in May. The lowest flows occur usually in the autumn months.

Natural lakes in the Carpathians are relatively rare, and all of them are quite small. Kondracki (2022) estimated the presence of some 450 lakes with a total surface area around 3.9 square kilometres. Lakes of glacial origin represent a substantial part of the natural lakes and are mainly located above the local treeline of previously glaciated massifs.

There are approximately 270 glacial lakes in the Western Carpathians (including small < 0.01 ha and/or seasonal ones) mostly in the High and Western Tatra Mountains, partly also in the Low Tatra Mountains. The largest among them (surface area almost 35 ha, max. depth 50–79 m) are situated in the northern (Polish) side of the Tatra Mountains where glaciers were the most extensive.

In the Ukrainian part of the Eastern Carpathians, in the Svydovets and the Chornohora massifs, there are several dozen glacial lakes, all very small and shallow. Many of them are in an advanced stage of paludification and many are transformed already to peat bogs. The water surface area of the largest lakes reaches recently quite exceptionally nearly 2.6 ha (lake Dogyaska) and a maximum depth of 4.8 m (lake Maricheyka) (Mykitchak et al. 2014, Mykitchak 2016).

Around 270 glacial lakes, including glacial peat bogs and peat bogs with pool, are reported from the Romanian part of the Carpathians. Although they are also found in the highest massifs of the Eastern Carpathians (e.g., Maramures, Rodna), most of them (81%) are in the Southern Carpathians, especially in the Retezat, Fagaras and Parang Mountains. The largest among them is lake Bucura (8.9 ha) and the deepest one is lake Zanoaga (29 m) (Pop et al. 2012, Necsoiu et al. 2016).

Other natural lakes were created in mountain areas when valleys were dammed after landslides and rock collapses caused by catastrophic events (earthquakes, extreme rainstorms). These lakes can be found mainly in the Eastern Carpathians. The largest of them are Lacul Rosu (Romania) and Morské oko (Slovakia) with an area of 12–13 ha and maximum depth of 12.5 and 25 m, respectively; however, with an area of 7 ha and max. depth of 22 m, lake Synevir, Ukraine is also worth mentioning.

Lake Sfanta Ana (area 19 ha, max. depth 7 m) which is situated in a crater of an ancient volcano in the Ciomadul Hills forming the southernmost part of the volcanic region of the Calimani-Gurghiu-Harghita mountain range (Eastern Carpathians, Romania) has an exceptional position among the Carpathian lakes because of its volcanic origin (Gastescu 2010).

Some smaller lakes were formed in karst areas built of Mesozoic limestones. These lakes are rare, their total number is less than 20. One of the larger ones is Lake Ighiu situated in the Apuseni Carpathians (Romania) which covers over 5 ha with maximum depth 9 m (Mihaiescu et al. 2012).

The list of natural lake types can be expanded by adding a small number of floodplain lakes (usually abandoned meanders) and spring lakes situated in the Carpathian valleys (Rubín et al. 1987, Gastescu 2010).

In addition, there are many man-made lakes in the Carpathians. Besides some 50 storage basins (Kondracki 2022), many small reservoirs can be found in the region. They were built for various purposes, such as waterpower source, water supply, fish farming, transporting logs, recreation. Reservoirs built for mining purposes are particularly valuable. Some began to be built since Roman times (in the Apuseni Carpathians), but mostly in the 16th–18th centuries, and some, constructed in the 18th century in Štiavnické vrchy Mountains (Western Carpathians, Slovakia), have been preserved to this day (Gastescu 2010, Bitušík et al. 2018).

1.2 The Family Chironomidae

The Chironomidae family (Insecta, Diptera), commonly called "non-biting midges" as adults and "bloodworms" as larvae (Epler 1995), represents the most numerous and species-rich group of aquatic insects. More than 4,100 species have been described globally (Ferrington 2008), although the total number of species is estimated at up to 15,000 (Cranston 1995). They are distributed worldwide, occupying all types freshwaters, but some species live in brackish, terrestrial and marine environments (Cranston 1995; Sæther and Ekrem 2003). They often dominate aquatic insect assemblages in number of species and individuals (~80–100, sometimes more species per site) (Ferrington 2008). Due to their high species diversity and the presence in most aquatic habitats, the preimaginal stages of chironomids are used as an important part of biomonitoring and nature conservation (for more information Nicacio & Juen 2015). Subfossil chironomids have been among the most frequently used palaeoindicators of past environmental changes owing to their chitinized head capsules preserved in limnic sediments in relative high numbers for up to several thousand years (Langdon et al. 2010). The identification is often possible to genus or species level (Brodersen & Quinlan 2006; Brooks et al. 2007). They are ubiquitous and their distribution and abundance is influenced by environmental variables, such as temperature, water depth, amount of dissolved oxygen, pH, type of substrate, productivity or pollution (Armitage 1995; Brooks et al. 2007), therefore reveal the environmental conditions at the time of sedimentation (Brodersen & Quinlan 2006). Changes in one or more of these factors are often accompanied by changes in the entire chironomid community. The chironomid-temperature relationship used in numerical chironomid-based inference models in the recent years enabled quantitative reconstructions of past temperature changes (Eggermont & Heiri 2012). Moreover, their analysis as a proxy helps to interpret the observed changes in multi-proxy studies, because each proxy provides different information about the studied ecosystem (Birks & Birks 2006).

1.3 A Brief Overview of Carpathian Chironomidae Research

In this brief review of the history of chironomid investigations in the Carpathians we focus on the countries where chironomids have been used as a proxy in palaeolimnological studies.

In comparison with West European countries with longer tradition of Chironomidae research, knowledge on the chironomid fauna in Slovakia, Poland and Romania is moderate, with the regional checklists consisting of 400–517 species (Bitušík et al. in prep., Siciński 2007, Tatole 2023). Moreover, the about 300 species recorded in Ukraine (Baranov 2011) indicate poor knowledge of the family in this country.

In the Chironomidae research of the Western Carpathians, the Tatra Mts. play a central role. Lakes of glacial origin has particularly attracted the attention of naturalists for their unique characteristics such as low temperatures, short ice-free period, low nutrient content, very little human impact, and consequently, exceptional fauna and flora. Because of these features, researchers recognized lakes as excellent contrasting biotopes to lakes and ponds situated in the lowlands strongly affected by human activities.

The beginning of chironomid survey of the Tatra Mts. dates back to the second half of the 19th century, when Nowicki (1873a,b) listed a number of species from this area for the first time and described three new species for science, including *Pseudodiamesa branickii*, which is still a valid species.

After a longer break, further research focused on a classification of lakes using chironomid and oligochaete assemblages (Hrabě 1939, 1942) with a description of some new species, such as, e.g., *Zalutschia tatrica* and *Procladius tatrensis*.

Not until the initiative of prof. Karol Starmach (Poland) and Dr. Milan Ertl (Slovakia) in the 1960s, started a complex limnological research on the biota of the Tatra lakes and streams. Chironomids, as the most important part of the benthic macrofauna, were initially investigated (thanks to Eva Ertlová, Marta Kownacka and Andrzej Kownacki) as indicators of changes in the trophic state of some lakes and, since the 1980s, as indicators of their acidification (for more details see Bitušík et al. 2006b, Kownacki 2010), and recently as indicators of recovery from acidification and changes caused by climatic change. It should be mentioned that during the limnological survey of the Tatra Mts. lakes, many new chironomid species were described, such as *Syndiamesa serratosioi*,

Diamesa starmachi, *D. nowickiana*, and *Chaetocladius bitusiki*; also, larval and pupal stages of previously known species (e.g., *Nilotanypus dubius* and *Parametriocnemus borealpinus*) were described (Kownacki 2010, Moubayed & Bitušík 2019).

The Slovak palaeolimnological research started in the late 1990s with Slovakia's participation in large international EU projects, such as AL:PE, MOLAR and EMERGE. Detailed analysis of the present research shows that ca. 75% of the Western Carpathian palaeolimnological studies, in which chironomids were used as palaeoindicators, were performed in the Tatra Mts. (Stoklasa et al. 2017). It is due to the high number of lakes without direct human impact and ability of the mountain region to react rapidly and sensitively to environmental changes that are preserved in lake sediments. So far, only two such studies have been realized outside the Tatra Mts.: in the Štiavnické vrchy Mts. (Bitušík et al. 2018) and in the Low Tatra Mts. (Jambrović 2022).

Overall, most of the chironomid palaeolimnological studies in the Western Carpathians has been focusing on a simple ecological reconstruction of the past environment and specific processes, such as acidification, eutrophication and temperature oscillation (see Klapyta et al. 2016, Stoklasa et al. 2017 for review). The only quantitative study trying to reconstruct temperature changes using the chironomid-summer air temperature inference model was conducted in the Slovak part of the East Carpathians (Hájková et al. 2016).

The beginning of chironomid research in Romania early in 1950s relates to prof. Nicholas Botnariuc, who, in cooperation with Victoria Cure, Paula Albu and Victoria Tatole, dealt with taxonomy and ecology of Chironomidae. Alike the Western Carpathians, the highest massifs in the Southern Carpathians also attracted the attention of researchers in Romania. Especially, chironomids from stagnant and flowing waters in the Retezat Mts. were studied systematically for more than 30 years (for review see Botnariuc & Tatole 1997). As a result, 117 chironomid species are known from this area including newly described species, *Bryophaenocladius pectinatus* and *Polypedilum* (*Tripodura*) *elongatum*, together with probably endemic *Georthocladius retezati* and *Heleniella extrema* (Cogalniceanu et al. 2009).

Palaeolimnological research in the Southern Carpathians started in the 2000s (Buczkó et al. 2009) and resulted in multi-proxy studies, that used chironomids primarily for temperature reconstructions or as a complementary proxy for the reconstruction of limnological change in lakes of the Ciomadul Mountains (Buczkó et al 2022, Magyari et al. 2022), Retezad Mountains (Tóth et al. 2012, 2018, 2022, Vincze et al. 2017), Rodna Mountains (Diaconu et al.

2017), Fagaras Mountains (Szabó et al. 2020), and Parang Mountains (Szabó et al. 2024).

From the limnological and palaeolimnological point of view, lakes and peat bogs situated in the Ukrainian part of Eastern Carpathians are seriously underinvestigated. Detailed analysis shows that data on the major groups of the benthic fauna (including Chironomidae) are missing (Mykitchak et al. 2014). This is evidenced by the results of a short chironomid survey of some alpine lakes, which showed that up to half of the found species were recorded on the territory of Ukraine for the first time (Bitušík et al. 2020). Alpine lakes represent a tiny fraction of all natural and artificial lakes in Ukraine in terms of number and size. Nevertheless, they are particularly suitable for studying ecosystem responses to both global (e.g., climate change, atmospheric pollution) and regional (e.g., land use change, species introductions) environmental impacts. Accordingly, they should be appreciated and studied more thoroughly in the future.

1.4 Study Lakes

A total of 72 lakes were surveyed in the Western and Eastern Carpathians (Fig. 1.2). In the Western Carpathians: Štiavnické vrchy Mountains, Low Tatra Mountains, Tatra Mountains (both in Slovakia and Poland), and Vihorlatské vrchy Mountains. In the Eastern Carpathians: the Chornohora and Svydovets Massifs in Ukraine, and the Maramures Mountains in Romania. Fig. 1.3 shows the distribution of the surveyed lakes along different gradients, such as altitude, area, depth, and pH.

1.4.1 *The Štiavnické vrchy Mountains*

Three reservoirs were studied in the Štiavnické vrchy Mts. in the inner belt of the West Carpathians that was formed during several stages of andesitic and rhyolitic volcanism in the Middle and Late Miocene (Chernyshev et al. 2013). The bedrock of the area consists of various types of andesite altered by successive hydrothermal activity (Konečný et al. 1998). The climate is slightly warm and humid with a mean annual air temperature of ~7.4 °C (at 600 m a.s.l.); the mean annual precipitation is 830 mm (Križáni et al. 2007).

The reservoirs are remnants of the former hydro-energetic system, which was established during the last period of the mining boom between the beginning of the 18th and the first half of the 19th century. Water from this system

FIGURE 1.2 Location of the surveyed lake districts within the Carpathians (indicated with grey): **1** Štiavnické vrchy Mountains (3 lakes), **2** Low Tatra Mountains (1 lake), **3** Tatra Mountains (52 lakes), **4** Vihorlatské vrchy Mountains (1 lake), **5** Chornohora and Svydovets Massifs (14 lakes), **6** Maramures Mountains (1 lake)

served as a hydraulic power source for mining and metallurgy purposes (see Chamutiová et al. 2018 for more details). Out of the almost 60 reservoirs present at the end of the 19th century, 30 have remained and serve mostly for recreation and angling.

The surveyed reservoirs, Rozgrund, Vindšachta and Veľká Richňava (Fig. 1.4), are located in an open mountainous landscape with a mosaic of mixed forests, grasslands and meadows. The reservoirs are thermally stratified, and their deepest parts are at the dam. Fish populations are present in all three reservoirs.

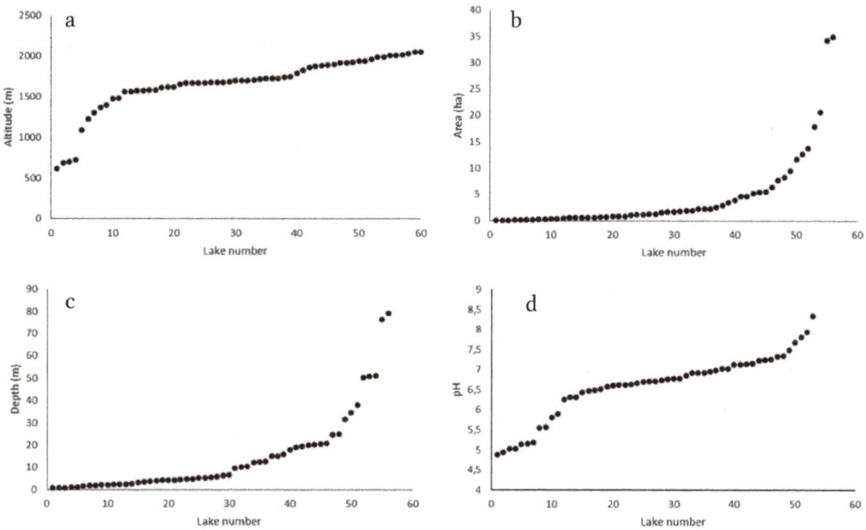

FIGURE 1.3 Distribution of the lakes in the training set along environmental gradients: **a** altitude, **b** lake surface area, **c** depth, and **d** pH

FIGURE 1.4 View of the reservoirs in the Štiavnické vrchy Mountains: **a** Rozgrund; **b** Vindšachta; **c** Veľká Richňava in its full state; **d** Veľká Richňava partly drained during the reconstruction of the dam
PHOTO P. BITUŠÍK (A–C) AND L. HAMERLÍK (D)

Recently, one of the reservoirs (Rozgrund) has become a drinking-water source with a protected watershed, while the other two became popular recreation sites with urbanization in their surroundings. For detailed description of the studied sites see Table 1.

1.4.2 *The Low Tatra Mountains*

The Low Tatra Mountains are the longest and second highest mountain range in Slovakia. The main ridge extending in the east-west direction has a length 95 km and exceeds the altitude of 2000 m in the central part (Ďumbier 2045 m a.s.l.).

Traces of the Pleistocene glaciation can be identified especially on the northern side of the mountain range in the form of glacial cirques, moraines and lakes. However, the glaciers were mostly small and thin, hence only eight small glacial lakes were formed at elevations ranging from ~1110 to 1680 m. The surveyed lake, Vrbické pleso, is the largest of them (Fig. 1.5).

In the study area, the crystalline bedrock consists mainly of granites and granodiorites with an admixture of various types of metamorphics (e.g., migmatites, gneisses). The climate is cool to cold, with average January temperatures of −5 to −7 °C and July temperatures of 11.5–16 °C (Turis & Jasík 2007). The surroundings of the lake are covered with a spruce forest (*Picea abies*) with

FIGURE 1.5 View of lake Vrbické pleso in the Low Tatra Mountains
PHOTO L. HAMERLÍK

an admixture of rowan (*Sorbus aucuparia*). The original peat bogs on the edge of the lake disappeared after the water level was artificially raised. At present, the lake catchment is impacted by the construction of infrastructure especially for winter activities.

1.4.3 *Tatra Mountains*

The Tatra Mountains, situated along the Slovak-Polish border in the Western Carpathians, are the highest and northernmost alpine-type range in the whole Carpathians with well-preserved glacial geomorphology of Pleistocene glaciations (Lukniš 1973). The Tatra Mts. also represent the most significant lake district in the Carpathians with 128 permanent lakes and about 123 periodic lakes of glacial origin. They are situated at elevations between 1089 and 2189 m a.s.l., and ~ 70% of them lie in the alpine zone above ~1800 m a.s.l. Generally, the lakes are relatively small, with areas of 0.01–34.5 ha and maximum depths of 0.5–79 m, and are frequently seeping, i.e., lack surface inflow and outflow.

The geology of the investigated area is based on crystalline bedrock. While the central part of the Tatra Mts. (High Tatra Mts.) is composed almost exclusively of granitoids, the western part (West Tatra Mts.) contains a significant amount of metamorphics (gneiss and mica schist) in addition to granodiorite.

The average annual air temperature is 1.6 and −3.8 °C at elevations of 1778 and 2635 m, respectively. The amount of precipitation increases with elevation, varying from ~1000 mm to ~1600 mm per year between 1330 and 2635 m a.s.l., but reaches > 2000 mm in some valleys. Precipitation is generally higher in the northern part than in the southern part of the mountains and snow cover usually lasts from October to June at elevations > 2000 m a.s.l.

The dominant vegetation in the catchments strongly depends on the altitude. Coniferous forest composed mostly of *Picea abies*, and partly of *Larix decidua* and *Pinus cembra*, grows below 1550 m a.s.l. The subalpine zone, situated between 1550 and 1800 m a.s.l., is characterized by dwarf pine (*Pinus mugo*) stands. Vegetation of the alpine zone (> 1800 m a.s.l.) is dominated by alpine meadows with patches of dwarf pine and an increasing proportion of rocks (see Bitušík et al. 2006; Stuchlík et al. 2017 for more details).

Fifty-two lakes were surveyed. The lakes are scattered along an altitudinal gradient from 1089 to 2157 m a.s.l. and represent all lake types from the forest through subalpine to alpine lakes (Figs. 1.6–1.9). The lakes are generally oligotrophic, but those in the forest zone are dystrophic (Figs. 1.6a–c). Some lakes were acidified in the second half of the 20th century. The lakes are mostly fishless, except for a few lakes in the West Tatra Mts., populated artificially with alpine bullhead (*Cottus poecilopus*), and in the central massif, populated

FIGURE 1.6 View of the Tatra Mountains forest- and subalpine lakes up to ~1600 m a.s.l.: **a** Vyšné Rakytovské pleso; **b** Smrekovické pleso; **c** Jamské pleso; **d** Popradské pleso; **e** Morskie Oko (on the left) and part of Czarny Staw pod Rysami (on the right); **f** Kolové pleso; **g** Veľké Biele pleso; **h** Dolné Roháčske pleso

PHOTO L. HAMERLÍK (A–C, E–G) AND P. BITUŠÍK (D,H)

FIGURE 1.7 View of the Tatra Mountains subalpine lakes situated up to ~1800 m a.s.l.: **a** Slavkovské pleso; **b** Vyšné and Nižne Temnosmrečinské pleso; **c** Malé Žabie Javorové pleso; **d** Trojrohé pleso; **e** Czarny Staw Polski, Wielki Staw and Przedni Staw; **f** Czarny Staw Gąsienicowy; **g** Vyšné Račkovo pleso, **h** Nižné Žabie Bielovodské pleso
PHOTO L. HAMERLÍK (A,B) AND P. BITUŠÍK (D–H)

FIGURE 1.8 View of the Tatra Mountains alpine lakes located up to ~2000 m a.s.l.: **a** Starolesnianské pleso; **b** Žabie Javorové pleso; **c** Nižné Terianske pleso; **d** Długi Staw (on the left) and Zadni Staw (on the right) Gąsienicowy; **e** Dlhé Velické pleso; **f** Litvorové pleso
PHOTO P. BITUŠÍK (A,B,D,F) AND L. HAMERLÍK (C,E)

(mostly artificially) with brook trout (*Salvelinus fontinalis*) and brown trout (*Salmo trutta*) (Figs. 1.6d, 1.7g). For further details on the studied lakes see Table 1.

1.4.4 *The Vihorlatské vrchy Mountains*

The Vihorlatské vrchy Mountains belong to the easternmost volcanic mountains in Slovakia. They are a part of a stratovolcano chain in the Carpathians

FIGURE 1.9 View of the Tatra Mountains alpine lakes above 2000 m a.s.l.: **a** Vyšné Wahlenbergovo pleso; **b** Ľadové pleso; **c** Pusté pleso; **d** Vyšné Terianske pleso; **e** Veľké and Prostredné Spišské pleso; **f** Veľké and Malé Žabie Mengusovské pleso
PHOTO L. HAMERLÍK (A) AND P. BITUŠÍK (B–F)

extending along the inner arch of the Eastern Carpathians from eastern Slovakia to northern Romania. The geology of the Vihorlatské vrchy Mts. was determined by volcanic activity in the Miocene (16–15 to 9.5 Ma). The bedrock of the lake catchment is based principally on several types of basaltic and pyroxene andesites (Pecskay et al. 2002).

The climate of the region is continental with mean annual temperatures of 4–6 °C (−5 to −6 °C in January, 14–16 °C in July) and with mean annual precipitation of 1000–1200 mm (http://geo.enviroportal.sk/atlassr). The catchment vegetation is dominated by dense beech forests.

FIGURE 1.10 View of lake Morské oko in the Vihorlatské vrchy Mountains. To estimate the size of the lake, note the floating platform in the middle of the lake (left picture)
PHOTO K. TRNKOVÁ (LEFT) AND L. HAMERLÍK (RIGHT)

Lakes in the area were formed during the Holocene (Terek et al. 2009, Pipík et al. 2022) by a landslide damming the valley into a tectonic depression.

Lake Morské oko (Fig. 1.10) is the third largest natural lake in Slovakia. However, the lake's current size was established in the 1890s, when an artificial dam was built, which caused an increase in surface area by 7 ha and in water level by 5 m. Morské oko is a dimictic lake considered to be mesotrophic based on phosphate concentration, transparency, and low oxygen concentration in the bottom (Terek 1981). During our investigation in 2017, the pH of lake water varied between 6.9 and 7.1. Majority of the lake bottom in the littoral consists of fine-grained sediment with a large amount of decaying vegetation from terrestrial input, while parts with steeply descending bottom are composed of coarse stones. Littoral macrophytes occur only around the mouth of the tributaries. Deep parts of the lake bottom are covered by gyttja (Terek 2009). The lake has five permanent and several periodical inflows and one outflow – the Okna brook.

Fish community consisting of small number of native species has been supplemented by the lake form of rainbow trout (*Oncorhynchus mykiss*) at the beginning of the 20th century when the native population of brown trout (*Salmo trutta*) decreased (Koščo et al. 2010).

1.4.5 *The Chornohora and Svydovets Massifs*

The Chornohora and Svydovets are the highest massifs of the Ukrainian part of the Eastern Carpathians. Both massifs show direct glacial imprints of the Pleistocene glaciations (Matoshko 2011), with lakes that belong to the most remarkable features of past glacier activity.

The climate is temperate with a moderate continental influence and varies significantly depending on topography. The July temperature varies from

20 °C at the southern margin of the mountains and 18 °C in the north to 6 °C on the highest peaks (Herenchuk 1968). Winter temperatures range from −3 °C to −10 °C, and mean annual precipitation reaches 1400 mm on the mountain ridges (Holubets et al. 1988).

Due to their small size (< 2 ha, max. depth < 2.2 m) and high elevation (above the upper tree line) all the study sites can be considered alpine ponds (Cereghino et al. 2008, Hamerlík et al. 2013). In fact, among the many lakes, only a few have an area larger than 0.1 ha and a maximum depth of more than

FIGURE 1.11 View of the lakes in the Ukrainian part of the Eastern Carpathians. Lakes in Chornohora: **a** Breskulets; **b** Breskul; **c** Nesamovyte; **d** Rudishina (Dantsyzh). Lakes in Svydovets: **e** Dogyaska; **f** Vorozheska (in the foreground) and Mala Vorozheska (in the background)
PHOTO L. HAMERLÍK

1.5 m (Mykitchak et al. 2014, Mykitchak 2016). Recently, most of them are in the advanced terrestrial phase, and many have already turned to peat bogs.

In the study area, bedrock is represented by sedimentary rocks of cretaceous paleogene flysch. Generally, the content of calcium in soils and their pH value in alpine habitats are higher in the Svydovets than in the Chornohora Mts. (Kobiv et al. 2009). The dominant vegetation of the lake catchment areas is formed by a unique mountain grassland ("polonyna") chequered, to various extent, by juniper (*Juniperus communis nana*), dwarf pine (*Pinus mugo*) or rhododendron (*Rhododendron* spp.) patches at some lakes.

A total of 14 lakes located between 1385 and 1793 m a.s.l. were studied (Fig. 1.11). The habitat structure of lake bottoms varies from stony silt to mud and organic depositions. Basic characteristics of the studied lakes are presented in Table 1. Lake names are given according to Mykitchak et al. (2014) and Mykitchak (2016).

1.4.6 *The Maramures Mountains*

The Maramures Mountains, Romania, is situated in the central part of the Carpathians by the Romanian-Ukrainian border. The mountain range is a part of the Eastern Carpathians and a continuation of the Ukrainian Chornohora Massif. The main ridge has a length of over 100 km, a total area of about 1500 square kilometres, and the elevation exceeds 1900 m a.s.l. Traces of the Pleistocene glacial activity, including lakes, are identified on the highest massifs – Pietrosul Maramuresului (Bardau) and Farcau-Mihailecu.

During the last glaciation, a few lakes were formed and by now most of them has completely turned into peat bogs. The biggest of them, our study site, lake Vinderel (Fig. 1.12) is situated at the foot of the Farcau peak (1957 m a.s.l.). The local geology comprises black flysch penetrated by basalt tuffs and basalt (https://muntiimaramuresului.ro/index.php/en/despre-2/geomorfologie). According to the measurements from the nearest meteorological station in the Rodna Mountain (1785 m a.s.l.), the sampling site has a subalpine climate characterized by low mean annual temperatures (0.6 °C) and precipitation of ~1300 mm per year (Farcas et al. 2013). Vegetation around the lake consists of alpine meadows with sparse clumps of *Pinus mugo, Juniperus nana* and *Picea abies*. Because of intense grazing, the area is marked with traces of gully erosion and shallow landslides (Balteanu et al. 2016).

FIGURE 1.12 View of lake Vinderel in the Maramures Mountains with the Farcau peak in the background
PHOTO L. HAMERLÍK

TABLE 1 Basic characteristics of the studied lakes and their subfossil chironomid assemblages. Lakes are grouped according to orographic units

Lake name/ orographic unit	**Alt.** m asl	**Coordinates**	**Lake area** ha	**Max depth** m	**pH**	**No. taxa**	**No. head capsules**	**Country**
Western Carpathians								
Štiavnické vrchy Mountains								
Rozgrund	703	48°28′39″N 18°52′32″E	5.44	20.9	–	12	31	SK
Veľká Richňava	725	48°25′37″N 18°50′46″E	8.28	19.5	–	10	24	SK
Vindšachta	688	48°26′03″N 18°51′22″E	4.65	12.7	–	17	46	SK
Low Tatra Mountains								
Vrbické pleso	1113	48°58′14″N 19°34′35″E	0.73	8	7.45	7	72	SK
Tatra Mountains								
Batizovské pleso	1884	49°09′08.3″N 20°07′53.4″E	3.48	10.5	6.43	5	132	SK
Čierne Kežmarské pleso	1579	49°12′31.92″N 20°13′25.45″E	0.29	4.0	7.24	5	38	SK
Dlhé Velické pleso	1939	49°09′54.87″N 20°08′40.63″E	0.63	5.6	6.32	4	98	SK
Dolné Roháčske pleso	1562	49°12′23.13″N 19°44′42.61″E	–	–	–	10	47	SK

TABLE 1 Basic characteristics of the studied lakes and their subfossil chironomid assemblages (*cont.*)

Lake name/ orographic unit	Alt. m asl	Coordinates	Lake area ha	Max depth m	pH	No. taxa	No. head capsules	Country
Dračie pleso	2020	49 °09′57.30″N 20 °05′18.60″E	1.72	16.0	6.93	5	105	SK
Horné Roháčske pleso	1728	49 °12′24.14″N 19 °44′06.23″E	–	–	–	8	115	SK
Kolové pleso	1565	49 °13′15.28″N 20 °11′32.37″E	1.83	1.2	6.78	16	96	SK
Ľadové pleso	2057	49 °11′2.65″N, 20 °9′41.01″E	1.70	18.0	6.62	3	41	SK
Ľadové pleso v Zlomiskách	1925	49 °09′47.3″N 20 °06′19.33″E	2.25	9.7	6.70	7	98	SK
Litvorové pleso	1860	49 °10′37.20″N, 20 °7′44.83″E	1.86	19.1	6.98	10	82	SK
Malé Hincovo pleso	1921	49 °10′26.4″N 20 °03′30.6″E	2.20	6,4	7.26	7	60	SK
Malé Žabie pleso	1920	49 °10′23.82″N 20 °04′28.46″E	1.21	12.6	6.67	8	113	SK
Malé Žabie Javorové pleso	1704	49 °12′9.06″N, 20 °8′58.05″E	0.18	3.1	6.95	13	306	SK
Nižné Kozie pleso	1942	49 °09′48.7″N 20 °02′36.3″E	0.78	2.3	6.92	6	86	SK
Nižné Rakytovské pleso	1307	49 °07′29.48″N 20 °01′33.57″E	0.13	2.1	5.3	16	104	SK
Nižné Rumanovo pleso	2094	49 °10′10.24″N 20 °06′01.60″E	–	–	–	3	102	SK
Nižné Spišské pleso	1993	49 °11′24.17″N 20 °11′50.22″E	0.62	4.3	6.26	7	97	SK
Nižné Temnosmrečinské pleso	1677	49 °11′34.4″N 20 °01′50.2″E	11.70	38.1	7.33	6	287	SK
Nižné Žabie Bielovodské pleso	1675	49 °12′01.02″N 20 °05′41.44″E	4.68	20.5	6.60	12	106	SK
Prostredné Sivé pleso	2013	49 °11′2.67″N 20 °10′31.01″E	1.8	4.8	6.31	9	96	SK
Prostredné Zbojnícke pleso	1960	49 °10′42.03″N 20 °9′38.01″E	0.60	5.3	6.76	8	78	SK
Pusté pleso	2056	49 °10′55.52″N 20 °9′14.02″E	1.19	6.6	6.92	10	161	SK
Slavkovské pleso	1676	49 °09′09.0″N 20 °11′04.9″E	0.11	2.5	5.19	11	191	SK
Starolesnianske pleso	1988	49 °10′48.0″N 20 °10′04.1″E	0.72	4.2	5.14	10	152	SK

TABLE 1 Basic characteristics of the studied lakes and their subfossil chironomid assemblages (*cont.*)

Lake name/ orographic unit	Alt. m asl	Coordinates	Lake area ha	Max depth m	pH	No. taxa	No. head capsules	Country
Strelecké pleso	2036	49 °11′02.67″N 20 °11′00.39″E	–	–	–	8	103	SK
Studené pleso	1825	49 °10′47.21″N 20 °10′34.86″E	–	–	–	12	98	SK
Velické pleso	1666	49 °09′24.03″N 20 °09′22.95″E	2.23	4.6	6.62	15	99	SK
Veľké Biele pleso	1615	49 °13′17.09″N 20 °13′50.18″E	0.97	0.8	6.69	19	150	SK
Veľké Spišské pleso	2013	49 °11′33.13″N 20 °11′41.31″E	2.87	10.1	6.48	9	95	SK
Vyšné Furkotské pleso	1698	49 °8′36.76″N 20 °1′47.90″E	0.41	2.4	6.52	6	57	SK
Vyšné Račkove pleso	1697	49 °11′59.17″N 19 °48′16.09″E	0.74	12.3	7.49	23	103	SK
Vyšné Satanie pliesko	1894	49 °10′12.7″N 20 °03′45.7″E	0.20	3.5	5.16	8	127	SK
Vyšné Smrekovické pleso	1368	49 °07′31.29″N 20 °02′06.06″E	–	–	–	15	106	SK
Vyšné Temnosmrečinské	1725	49 °11′20.7″N 20 °02′22.2″E	5.56	20.0	7.35	10	190	SK
Vyšné Terianske pleso	2124	49 °10′04.8″N 20 °01′18.5″E	0.56	4.3	5.55	7	43	SK
Vyšné Wahlen-bergovo pleso	2157	49 °09′51.1″N 20 °01′37.6″E	5.17	20.6	6.58	9	27	SK
Vyšné Žabie Bielovodské pl.	1699	49 °11′39.1″N 20 °05′39.5″E	9.46	24.8	6.64	9	103	SK
Zelené Kačacie pleso	1575	49 °10′38.7″N 20 °06′59.23″E	2.53	2.7	7.13	16	121	SK
Žabie Javorové pleso	1878	49 °11′28.89″N 20 °10′7.33″E	1.13	15.3	7.16	8	124	SK
Czarny Staw Gąsienicowy	1620	49 °13′56.35″N 20 °01′01.7″E	17.94	51.0	6.49	14	106	PL
Czarny Staw pod Rysami	1580	49 °11′19.7″N 20 °04′40.1″E	20.64	76.4	7.14	20	58	PL
Czarny Staw Polski	1722	49 °12′18.32″N 20 °01′46.44″E	12.69	50.4	6.73	12	69	PL
Kurtkowiec	1686	49 °13′48.03″N 20 °00′07.56″E	1.53	4.8	6.70	15	85	PL
Litworowy Staw	1622	49 °14′01.20″N 19 °59′49.23″E	–	–	–	26	93	PL

TABLE 1 Basic characteristics of the studied lakes and their subfossil chironomid assemblages (*cont.*)

Lake name/ orographic unit	Alt. m asl	Coordinates	Lake area ha	Max depth m	pH	No. taxa	No. head capsules	Country
Morskie Oko	1395	49°11'50.0"N 20°04'12.7"E	34.93	50.8	7.13	27	87	PL
Przedni Staw Polski	1668	49°12'44.0"N 20°02'54.7"E	7.71	34.6	7.23	17	72	PL
Smreczynsky Staw	1226	49°13'19.02"N 19°51'48.85"E	0.75	5.3	4.95	12	82	PL
Toporowy Staw Nižny	1089	49°17'02.00"N 20°01'50.37"E	0.62	5.7	5.57	18	43	PL
Wielki Staw Polski	1655	49°12'32.9"N 20°02'22.6"E	34.35	79.3	7.30	18	46	PL
Zadni Staw Polski	1890	49°12'46.6"N 20°00'47.0"E	6.47	31.6	6.78	7	90	PL
Zielony Staw Gąsienicowy	1672	49°13'45.26"N 19°59'51.68"E	3.84	15.1	6.85	15	93	PL
Zmarzly Staw Gąsienicowy	1787	49°13'26.69"N 20°1'20.44"E	0.30	3.7	5.90	10	109	PL
Eastern Carpathians								
Vihorlatské vrchy Mountains								
Morské oko	619	48°54'51"N 22°11'56"E	13.8	25.1	7.2	9	89	SK
Chornohora Massif								
Nesamovyte	1745	48°7'20.6"N 24°32'22"E	0.35	2.0	5.81	9	106	UA
Breskulets	1738	48°8'57.9"N 24°30'13.3"E	0.04	1.1	5.30	11	95	UA
Breskul	1728	48°8'53.3"N 24°30'14.8"E	0.01	1.6	4.89	6	93	UA
Rudishina (Dantsyzh)	1671	48°7'52.6"N 24°32'13.7"E	0.05	0.9	7.81	19	111	UA
Brespo	1627	48°08'37.4"N 24°30'56.1"E	0.05	0.5	7.35	5	56	UA
Brebenescul	1793	48°06'06.0"N 24°33'44.2"E	0.61	3.2	7.86	9	68	UA

TABLE 1 Basic characteristics of the studied lakes and their subfossil chironomid assemblages (*cont.*)

Lake name/ orographic unit	Alt. m asl	Coordinates	Lake area ha	Max depth m	pH	No. taxa	No. head capsules	Country
Svydovets Massif								
Dogyaska	1584	48°16′11.2″N 24°9′55.1″E	1.90	2.0	7.68	7	87	UA
Vorozheska	1480	48°16′34″N 24°11′33.9″E	0.54	2.2	7.93	10	74	UA
Mala Vorozheska	1477	48°16′38.9″N 24°11′33.7″E	0.12	0.8	8.34	13	92	UA
Ivor 1	1600	48°13′42.0″N 24°14′06.1″E	0.13	1	6.29	7	65	UA
Ivor 2	1600	48°13′41.6″N 24°14′04.4″E	0.04	0.6	5.74	10	76	UA
Apshynets	1491	48°16′54.8″N 24°09′31.7″E	1.13	1.8	8.10	23	59	UA
Kosivske 1	1620	48°15′46.5″N 24°11′57.5″E	0.09	1	7.46	13	66	UA
Dragobratske	1385	48°14′26.7″N 24°14′26.2″E	0.07	1.5	8.48	18	65	UA
Maramures Mountains								
Vinderel	1684	47°54′34.7″N 24°27′19.6″E	0.90	5.5	–	11	91	RO

1.5 Key to Subfamilies/Tribes of Subfossil Chironomidae

The following key to subfamilies uses only characters usually present in subfossil remains and will distinguish most of the taxa. However, it may not separate some taxa and thus the key should be used with caution. We suggest combining it (especially for beginners) with other subfamily keys, e.g., Brooks et al. (2007) and Andersen et al. (2013). Subfamilies are written with capital letters. On the right from the subfamily/tribe name, a head capsule of a typical representant is figured. The complete list of morphotypes and their distribution within the studied lake districts is summarized in Appendix 1.

1 Mentum and ventromental plates absent, characteristic toothed structure (ligula) present instead; antenna, if present, usually retracted in head capsule, with long basal segment .... TANYPODINAE

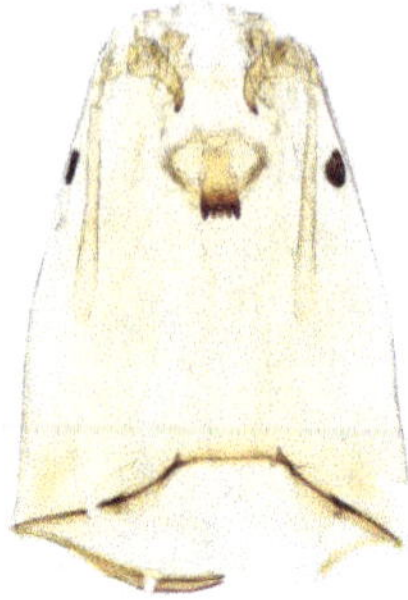

– Mentum strongly developed, with or without ventromental plates; antenna, if present, not retractile 2

2 Mentum expanded laterally to well-developed ventromental plates that are usually striated and always without beard. CHIRONOMINAE 3

– Mentum without ventromental plates, or, if plates present, they are never striated and sometimes with beard 4

3 Mentum with large, fan-shaped ventromental plates, widely separated medially, antennal pedestal small, without large spur .... tribe Chironomini

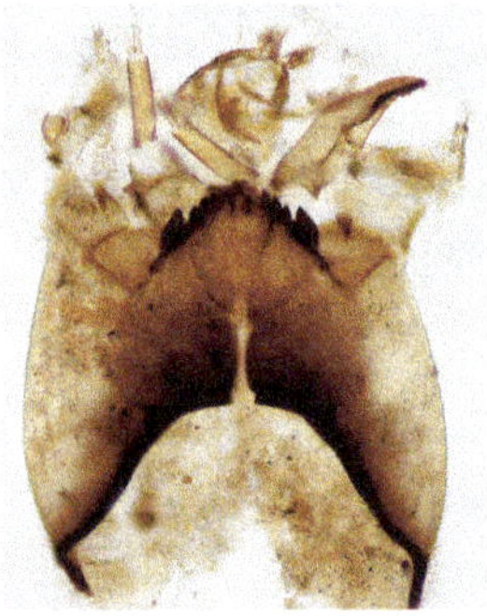

– Mentum with narrow ventromental plates, usually close to each other medially, antennal pedestal prominent, with or without spur, antennae, if present, usually very long tribes .... Tanytarsini, Pseudochironomini

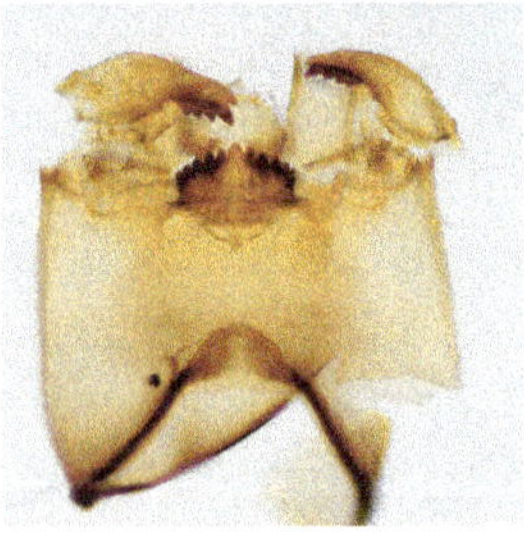

4 Mentum usually with more than 6 pairs of lateral teeth; antenna with 3rd segment annulated 5

– Mentum usually with up to 6 pairs of lateral teeth, 3rd antennal segment normal, not annulated 6

5 Head usually strongly pigmented, with dark thick occipital margin .... DIAMESINAE

– Head less pigmented, yellow; occipital margin thin .... PODONOMINAE

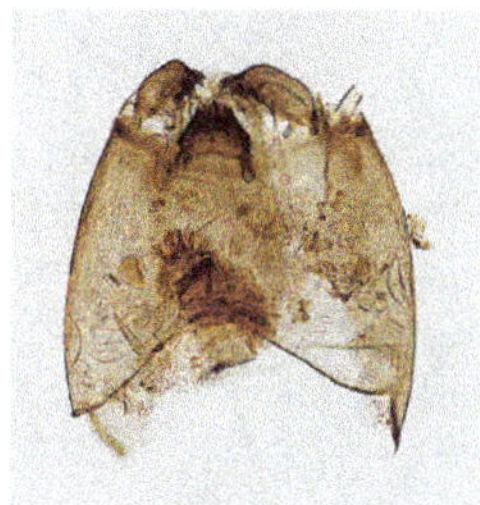

6 Ventromental plates large, without striation, strong beard present, usually reaching beyond lateral margin of head capsule .... PRODIAMESINAE

– Ventromental plates, if present, weak; if distinctive, then leaf- or blade-shaped; if dense beard is present, it is short, never reaching beyond lateral margin of head capsule .... ORTHOCLADIINAE

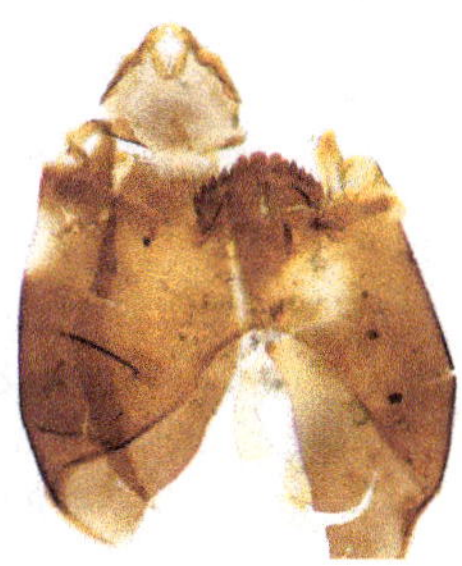

CHAPTER 2

Tanypodinae

Head capsules of the subfamily Tanypodinae can be distinguished from the other subfamilies at first glance. The most obvious feature, unique for this subfamily, is the presence of a ligula and a pair of paraligulae on either side of the base of the ligula. Another typical feature is the absence of distinct mentum and ventromental plates. If a toothed dorsomentum is present, it is less noticeable and of a different shape than in other subfamilies (Fig. 2.1). The mandible is curved, tapered towards apex into a dark apical tooth with one inner tooth

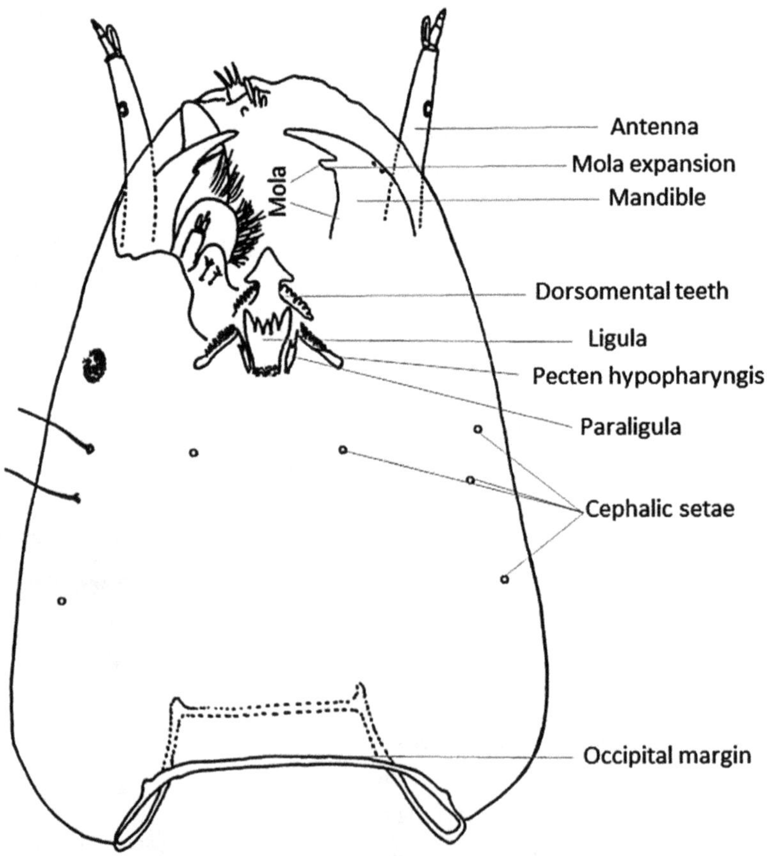

FIGURE 2.1 Tanypodinae larval head capsule showing subfossil diagnostic characters, comparing recent (left side) and subfossil remain (right side). Modified after Bitušík (2000)

 | DOI:10.1163/9789004694781_003

(rarely more). In some species the distal part of mola is swollen into a lobe appearing as a large tooth (in older literature called also basal tooth, Fittkau and Roback, 1983). Features used to distinguish morphotypes are mainly number of teeth, shape and color of the ligula, presence/absence of dorsomental teeth and the shape and color of occipital margin. In the subfossil material, the ligula and frequently also other structures important for identification, may be missing. In that case, the position of cephalic setae and sensory pores can be used for generic diagnoses (for details see Kowalyk 1985, Rieradevall & Brooks 2001, Brooks et al. 2007, Cranston & Epler 2013). Tanypodinae remains are less well preserved and considered underrepresented in fossil assemblages relative to other subfamilies (Walker et al. 1984, Walker 1987).

KEY TO GENERA OF TANYPODINAE

1 Ligula with 4 subequal teeth .... *Derotanypus*
– Ligula with 5 teeth 2

2 Head rounded, dorsomental teeth distinct 3
– Head prolonged, at least 2× as long as broad, dorsomental teeth absent 4

3 Apical half of ligula black, mandible with large blunt molar expansion; paraligula multitoothed .... *Procladius*
– Ligula unicolored, mola with a small pointy tooth at most, paraligula bifid .... *Macropelopia*

4 Apices of ligula teeth form a concave arc; basal segment of maxillary palp divided in 2–6 segments .... *Ablabesmyia*
– Ligula teeth subequal, forming a straight line; basal segment of maxillary palp undivided .... *Zavrelimyia*

Ablabesmyia Johannsen (Fig. 2.2)
Head capsule narrow, yellowish-brownish (Fig. 2.2a). **Occipital margin** dark with two thorn-like projections (Fig. 2.2f). **Mandible** with single long apical tooth with dark tip and bluntly rounded inner tooth, mola protruding as large blunt or pointed tooth. Basal segment of **maxillary palp** subdivided into 2–6 segments (in subfossil material often 2–3 segments are visible) (Figs. 2.2b,c). **Dorsomental** teeth absent. **Ligula** with 5 dark teeth, median tooth is the shortest, other gradually increase in size (Fig. 2.2e). **Paraligula** bifid, pecten hypopharyngis consists of up to 20 teeth of unequal size (Fig. 2.2d). Note, however, that larvae of the third instar have pecten with 6–8 teeth of equal size.

Remarks

Especially important diagnostic features to distinguish heads of mature larvae from other genera are segmented maxillary palps (Figs. 2.2b,c; except for *Paramerina*, currently a subgenus of *Zavrelimyia*, that has 2-segmented maxillary palp), thorn-like projections on occipital margin (Fig. 2.2f; except for some *Zavrelimyia*), and unequally sized pecten teeth (Fig. 2.2d).

Morphotypes

Subfossil morphotypes can be distinguished only when maxillary palps are present on the head capsule.

1 Basal segment of maxillary palp divided to 5–6 segments .... *Ablabesmyia longistyla*-type
– Basal segment of maxillary palp divided to 2 segments (a small 3rd segment may be present at base) .... *Ablabesmyia monilis*-type

Note that preimaginal stages of the two types (or the species they represent) are extremely similar (also pupal exuviae) and can be easily mismatched. Moreover, as Rieradevall & Brooks (2006) pointed out, while correctly stated in Laville (1971), the captions showing maxillary palps of the two species have been swapped in Fittkau & Roback (1983) and later in Cranston & Epler (2013). As a consequence, part of the larval material may be misidentified and, in turn, ecological requirements of both species/morphotypes and their distribution in waterbody types may be biased.

Distribution

Remains of the genus were common in all studied lake districts in SK and PL and in one lake in the Svydovets Massif (UA). In the Tatra Mts., *Ablabesmyia* was restricted to nutrient rich dystrophic lakes located in the coniferous forest zone.

Ecology and palaeoecological interpretation

Larvae of *Ablabesmyia* are eurythermic and eurytopic, occurring in a variety of standing waters, even in slow flowing rivers and canals. They live mainly among plants, and on sediments containing organic particles. Larvae cannot survive anaerobic conditions and are indifferent to pH. Due to their predatory habit, density of the larvae is usually low (Vallenduuk & Moller Pillot 2007, Bitušík & Hamerlík 2014).

FIGURE 2.2 *Ablabesmyia*. **a** head capsule, **b** maxillary palp (*Ablabesmyia monilis*-type), **c** maxillary palp (*Ablabesmyia longistyla*-type), **d** pecten hypopharyngis, **e** ligula, **f** occipital margin with two thorn-like projections. (Figs. b and c are modified after Bitušík and Hamerlík 2014)

In the stratigraphic record, *Ablabesmyia* remains may indicate warmer conditions with sufficient content of dissolved oxygen, and usually higher (mesotrophy) trophic status (Heiri et al. 2003). Larvae can also be associated with macrophytes (Brodersen et al. 2001b).

Derotanypus Roback (Fig. 2.3)
Head capsule large, rounded oval, light orange-brown (Fig. 2.3a). **Occipital margin** dark brown. **Mandible** uniformly curved; mola with 2–6 adjacent saw-like teeth; several small teeth irregularly distributed on the surface of mandible in the apical half (Fig. 2.3d). **Dorsomentum** on each side with 4–7 teeth in a strongly concave row (Fig. 2.3c). **Ligula** with 4 teeth of equal size, tooth row straight. **Paraligula** large, up to ¾ lengths of ligula, outer margin with up to 10 slender branches (Fig. 2.3b). **Pecten hypopharyngis** with 15–30 subequal teeth.

Remarks

The ligula with 4 teeth (Fig. 2.3b) will distinguish *Derotanypus* from all other Palaearctic Tanypodinae, except for *Psectrotanypus*. However, *Psectrotanypus* larvae have pale head capsules (as opposed to orange-brown in *Derotanypus*),

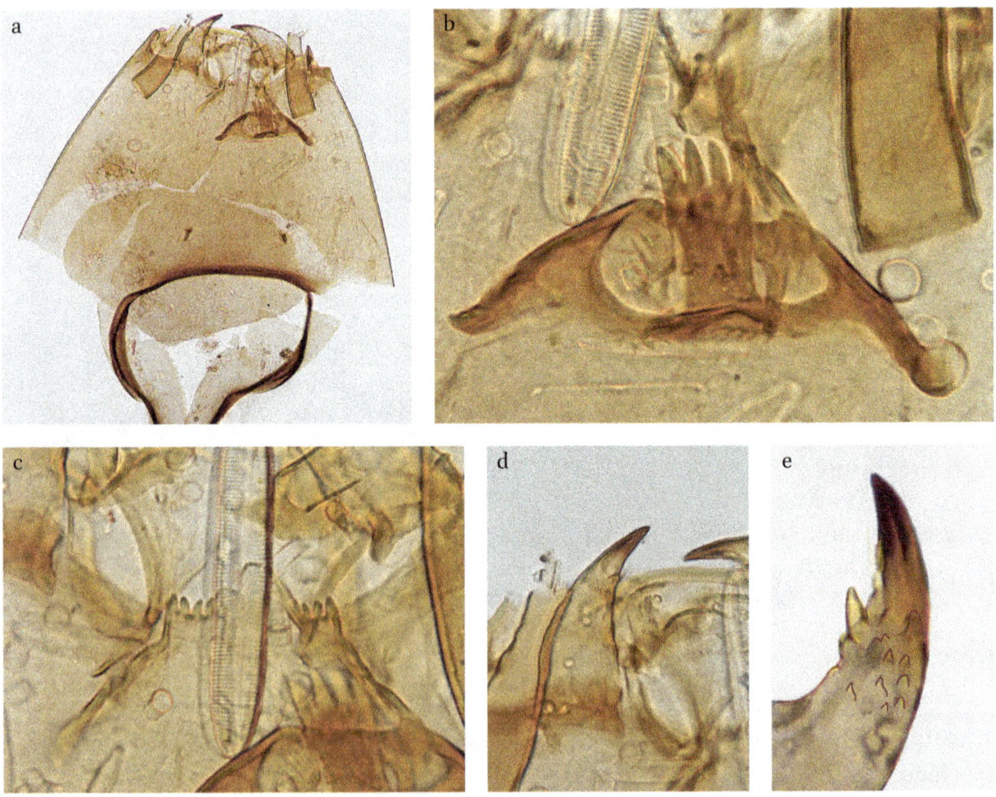

FIGURE 2.3 *Derotanypus* sp.: **a** head capsule, **b** ligula and paraligulae, **c** dorsomental teeth, **d** mandible, **e** detail of the apical part of mandible showing surficial teeth (modified after Bitušík and Hamerlík 2014)

straight dorsomentum (strongly biconcave in *Derotanypus*), and lack surficial teeth on the mandible (while irregularly distributed small surficial teeth are present in *Derotanypus*).

Morphotypes

The recorded head capsule was not assigned to any species/morphotype. However, there is only one species, *Derotanypus sibiricus* (Kruglova & Chernovskii 1940), know from Europe (Sæther & Spies 2004), thus it is likely that the recorded type belongs to this species.

Distribution

One head capsule was found in a dystrophic forest lake in the Tatra Mts. (SK).

Ecology and palaeoecological interpretation

Derotanypus larvae are oligostenothermic and live in silty substrates of small waterbodies, such as cold ponds and streams in the northern and mountainous areas of the Holarctic (Cranston & Epler 2013). Novikmec et al. (2015) found living larvae in some subalpine Tatra ponds.

Derotanypus remains are scarce in lake sediments, perhaps also because they are associated with small waterbodies, which are used infrequently for palaeolimnological reconstructions. *Derotanypus* remains were found in late-glacial sediments from a carbonate lake in England (Brooks et al. 2007), or together with other oligotrophic and stenothermic chironomid taxa during the initial lake development phase (Allerød interstadial) of a peat bog in the Vihorlatské vrchy Mts., Slovakia (Hájková et al. 2016). In Vrbické pleso, Low Tatra Mts., *Derotanypus* was among the dominating morphotypes in sediment samples most likely representing the coldest period of the LIA (Jambrović 2022). In the palaeorecord *Derotanypus* remains will refer to cold conditions and small water volume.

Macropelopia Thienemann (Fig. 2.4)

Head capsule broad, yellowish to orange in life, subfossil heads reddish-brown (Fig. 2.4a). **Occipital margin** dark brown. **Mandible** slender, uniformly curved; mola with 1 distal and 1 or more marginal teeth, with small dorsolateral tooth (difficult to see from lateral view) (Figs. 2.4b,d). **Dorsomentum** with 7–9 teeth. **Ligula** with 5 teeth, tooth row deeply concave, outer tooth 2× as long as the median one (Figs. 2.4b,c). **Paraligula** bifid (Fig. 2.4c). **Pecten hypopharyngis** with 15–25 subequal teeth.

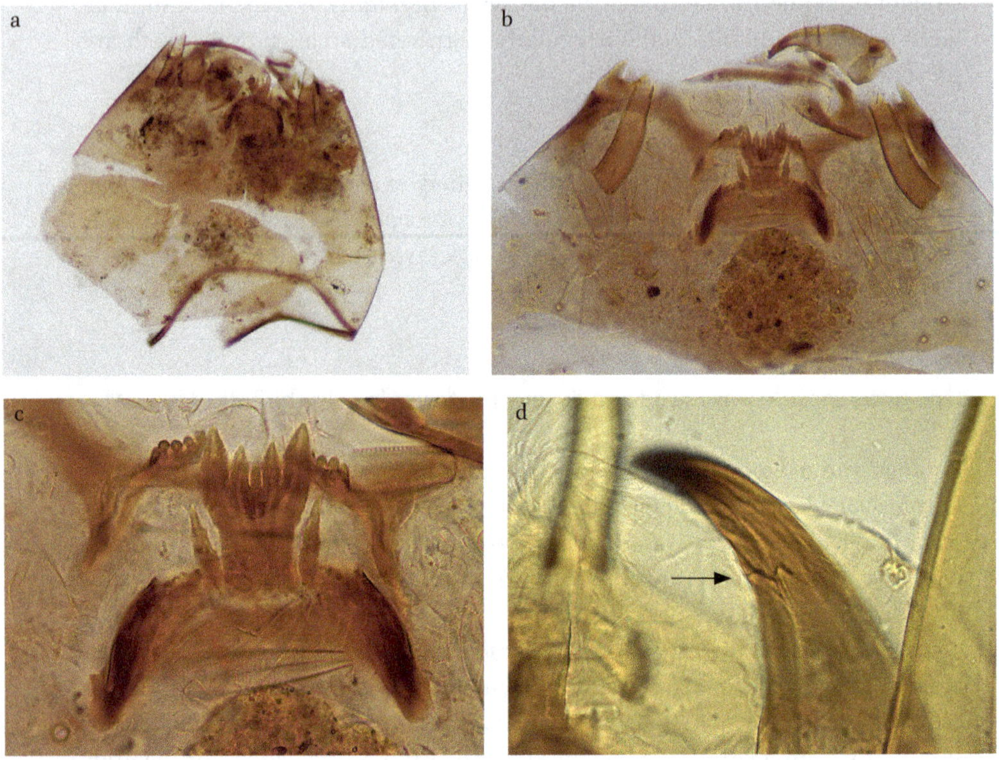

FIGURE 2.4 *Macropelopia nebulosa*-type: **a** head capsule, **b** ligula and mandible, **c** ligula, paraligula and dorsomental teeth, **d** detail of mandible (note the double tooth on the mola; modified after Bitušík and Hamerlík 2014)

Remarks

The related genus *Apsectrotanypus* (not shown in this book) differs from *Macropelopia* in having only 4 large and 1 smaller dorsomental teeth (note that younger larval instars of *Macropelopia* have reduced number of dorsomental teeth), and brown head with light area around eye spots and median pale stripe on ventral side of the head capsule.

Morphotypes

One morphotype was recognized.

***Macropelopia nebulosa*-type**. The type was suggested based on the shape of inner ligula teeth (Fig. 2.4c). While points of inner teeth are rather straight in larvae of this group, they are distinctly curved outwards in *M. notata* group (Cranston & Epler 2013).

Distribution

Subfossil remains were common (but never abundant) in all lake districts (SK, PL, UA) except for lakes Morské oko (SK) and Vinderel (RO). In the Tatra Mts. the genus occurred especially in lakes situated below 2000 m a.s.l.

All remains were morphologically identical and tentatively considered as one morphotype.

Ecology and palaeoecological interpretation

Fittkau (1962) considered the genus as slightly oligostenothermic, but recent data indicate wider temperature tolerance (Vallenduuk & Moller Pillot 2007 and references therein). The three most common European species, *M. nebulosa* (Meigen, 1804), *M. adaucta* Kieffer, 1916 and *M. notata* (Meigen, 1818), have notably different ecological requirements, thus here we refer to the ecology of the most common species in Slovakia, most likely also present in the subfossil material, *M. nebulosa*. However, it must be said that Michailova *et al.* (2014) recently reported the presence of *M. rossaroi* Lencioni & Marziali, 2005 in two lakes on the Polish side of the Tatra Mts., belonging to the *M. nebulosa* group.

Larvae occur in low densities in soft sediments (sand and detritus) of various types of stagnant waters without respect to altitude, and in slowly flowing lowland waters. In stagnant waters, larvae live in more or less stable oxygen content (Vallenduuk & Moller Pillot 2007, Bitušík & Hamerlík 2014).

Macropelopia remains are usually infrequent in stratigraphic records. The changes in relative abundance can be associated with changes in oxygen regime. Brooks et al. (2007 and citation therein) considered it as acidophilic, although our data from the Tatra Mts. lakes do not clearly confirm this (Bitušík et al. 2006). Like with other predacious tanypods (see comments to *Procladius*), changing pattern in the sediment sequence could be attributed to changes in prey abundance. Moreover, it seems that *Macropelopia* is negatively affected by fish predation (Raposeiro et al. 2017, Slobodníková et al. 2023).

Procladius Skuse (Fig. 2.5)

Large, oval **head** capsule whitish to yellowish in color (Fig. 2.5a). **Mandible** slender, uniformly curved with dark black tip of the apical tooth; mola distally expanded to a large, broad tooth with bluntly rounded apex (Fig. 2.5c). **Dorsomental** teeth (6–8) present. **Ligula** with 5 dark teeth, their apex is blackish brown. **Paraligula** multibranched with usually prominent main spine (Fig. 2.5b). Shape of paraligula is diagnostic to differ the subgenera *Holotanypus* and *Psilotanypus* (Cranston & Epler 2013). **Pecten hypopharyngis** with 10–15 teeth and some smaller teeth in a supplementary row.

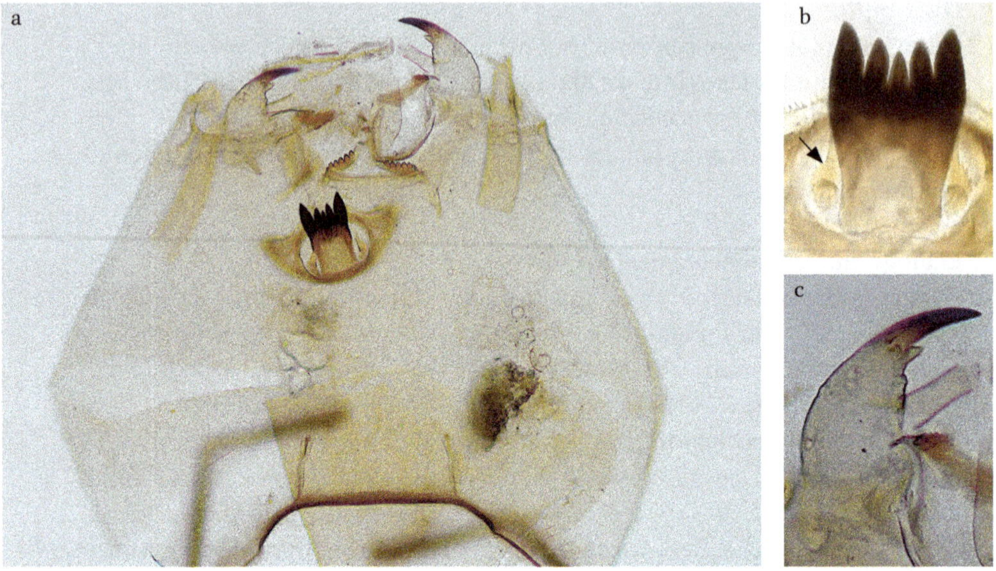

FIGURE 2.5 *Procladius*. **a** head capsule, **b** ligula (arrow indicates paraligula) (modified after Bitušík and Hamerlík 2014), **c** mandible

Remarks

Procladius remains differ from all other Tanypodinae with dorsomental teeth, in blackish-brown coloration of distal part of ligula (Fig. 2.5b) and apex of mandibula, and in large, blunt mandibular mola expansion (Fig. 2.5c). Morphological deformities of mouthparts (mainly in ligula and paraligulae) and antennae can occur in response to serious water contamination by PCBs and heavy metals (Warwick 1991 and references therein), however without effect on correct identification.

Morphotypes

One morphotype, *Procladius* sp. was identified.

Distribution

Subfossil remains were common in lakes of all lake districts. All remains belonged to subgenus *Holotanypus*.

Ecology and palaeoecological interpretation

Procladius larvae are typical bottom dwellers living in mineral and organic substrata, usually not on stones or plants. They can occur in stagnant and slow flowing waters regardless of size and trophic conditions (Vallenduuk &

Moller Pillot 2007, own observations). The larvae can move, occasionally or seasonally, from oxygen poor habitats to water with optimal oxygen content. *Procladius* seems to be indifferent to pH, for example in the Tatra Mts., the larvae were also found in strongly acidified lakes with pH < 5.0 (Bitušík et al. 2006). Generally, larval abundance is controlled by animal prey rather than abiotic conditions (Bitušík et al. 2010 and citation therein).

Even though many authors assign *Procladius* as a profundal species (see Brooks et al. 2007), larvae can be very abundant in the littoral zone, too. Moreover, studies on the vertical distribution of larval chironomids in some relatively deep lakes (up to 25 m depth in lakes Popradské pleso, Štrbské pleso and Morské oko) showed, that *Procladius* density peaked between 5 and 8 m depth and further down larvae disappeared or their number decreased significantly (unpublished data).

Procladius subfossils can be abundant in stratigraphic records of different lake types, and more detailed identification of head capsules would be very useful as several species show different lake type preference with respect to depth, trophy and temperature (Brodersen et al. 2001a, Engels et al. 2008). Unfortunately, such level of taxonomic identification of larval subfossils is not possible yet.

Generally, in thermally stratified lakes, the succession of chironomid communities towards *Chironomus* and *Procladius* assemblage indicate decrease in hypolimnetic oxygen (Brodersen et al. 2001a, Quinlan & Smol 2001, Dimitriadis & Cranston 2001), increase of a lake productivity (Brodersen & Quinlan 2006, Magny et al. 2006) and more mature lacustrine conditions (Alm & Willassen 1993). Some authors (e.g., Luoto & Nevalainen 2013) consider *Procladius* an indicator of warmer lake conditions; however, the broad ecological plasticity of this large genus should be emphasized. Larvae of different species inhabit the whole gradient from warm and productive lowland waterbodies to cold and unproductive alpine lakes (Heiri et al. 2003, Bitušík et al. 2006), and this fact should be taken into account in the palaeolimnological interpretation. Moreover, absence of *Procladius* remains (similar to other large-bodied, free living Tanypodinae) can be connected with the presence of salmonids in alpine lakes (Heiri & Lotter 2003, Slobodníková et al. 2023).

Zavrelimyia Fittkau (Fig. 2.6)

Head capsule narrow yellowish-brown, in some species partially dark in proximal part (Fig. 2.6a). **Occipital margin** darker with two thorn-like projections on ventral side (Fig. 2.6c). **Mandible** moderately curved with long apical tooth and large inner tooth. Mola expanded into large, blunt tooth (Fig. 2.6a).

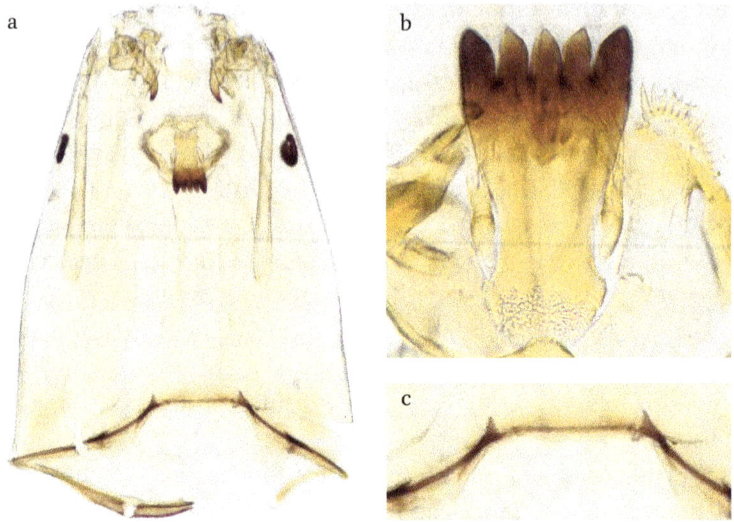

FIGURE 2.6 *Zavrelimyia*. **a** head capsule, **b** ligula, **c** occipital margin with two thorn-like projections. Modified after Bitušík and Hamerlík (2014)

Dorsomentum without teeth. **Ligula** with 5 brown subequal teeth (Fig. 2.6b). **Paraligula** bifid. **Pecten hypopharyngis** with 9–15 teeth, inner tooth and some middle teeth enlarged.

Remarks

Zavrelimyia resembles *Paramerina* (recently moved to *Zavrelimyia* as a subgenus, Silva and Ekrem, 2016) with similar arrangement of teeth on ligula, but basal segment of maxillary palp in *Paramerina* is subdivided into 2 parts. If this characteristic is absent from subfossil material, *Zavrelimyia* can be separated by having thorn-like projections on the occipital margin (Fig. 2.6c).

Morphotypes

One morphotype, *Zavrelimyia* sp. was distinguished.

Distribution

Zavrelimyia was the second most abundant tanypod remain, mostly occurring in the surface sediments of the subalpine Tatra Mts. lakes (SK, PL) and also present only in the Chornohora Massif (UA) and absent from other studied sites.

Ecology and palaeoecological interpretation

Zavrelimyia is a large genus, with many species with a wide range of ecological requirements, which may make straightforward interpretation based on larval remains (that are impossible to identify to morphotypes) rather difficult. Larvae inhabit a wide range of habitats, from springs and fast flowing mountain waters to lowland brooks, and small water bodies in wetlands and woody areas, even temporary pools (Vallenduuk & Moller Pillot 2007 and citation therein). Unlike the eurythermic subgenus *Paramerina*, larvae of *Zavrelimyia* are considered oligostenothermic (except for the *nubila* group) and they inhabit sediments in lake littorals and lower sections of streams in high altitude or high latitude regions (Cranston & Epler 2013). Larvae of *Zavrelimyia* are widespread and abundant in the alpine lakes of the Alps and the Tatra Mts. (Boggero et al. 2006, Bitušík et al. 2006). Despite its cold-stenothermic character, *Zavrelimyia* is more plastic thermally than the extremely cold-adapted *Pseudodiamesa* and *Micropsectra radialis*-type. Thus, in stratigraphic records, its higher proportion can relate to warmer climate and more productive conditions (e.g., Heiri et al. 2003, Ilyashuk et al. 2011), however, for correct interpretation it is important to consider the structure of the original community. If the predatory behaviour of larvae is considered, the change in *Zavrelimyia* remains can suggest more complex interactions in the lakes (Tatosová et al. 2006). Also see comment to *Procladius*.

CHAPTER 3

Diamesinae

Head capsules of the subfamily are yellowish brown to dark brown or black with broad black occipital margin. Surface of dorsal head mostly smooth, rarely covered with numerous short setae (*Protanypus*), or with tubercles (*Boreoheptagyia*). Third antennal segment annulated, except for *Protanypus*. Premandible with 1–16 teeth. Mandible with apical tooth and 4–5 inner teeth. Mentum variable in shape, with 4 (*Protanypus*) to 23 teeth, rarely without teeth (*Potthastia longimanus* group). Ventromental plates rudimentary to very large (*Pseudodiamesa*).

Larvae of most species are rheophilic to rheobiontic but may also occur in the littoral zone of cold oligotrophic lakes.

KEY TO GENERA OF SUBFOSSIL DIAMESINAE

1 Mentum with triangular median tooth separated from first lateral teeth by deep V-shaped notch .... *Pseudodiamesa*

– Mentum with differently shaped median tooth 2

2 Mentum dark unicolored (sometimes central region is slightly lighter); median tooth single or double, 7–11 lateral teeth present, 2nd lateral tooth usually not smaller than 3rd tooth .... *Diamesa*

– Mentum with obvious color pattern, 5 central teeth much lighter than remaining laterals; median tooth single, 6–7 lateral teeth present; median tooth and 1st laterals subequal in size, 2nd lateral tooth significantly smaller than 1st and 3rd tooth .... *Pseudokiefferiella**

*Larvae of *Pseudokiefferiella* are extremely similar to *Diamesa* and can be distinguished with certainty only using characters on the body.

Diamesa Meigen (Fig. 3.1)

Head capsule yellow to dark brown or brown-black with robust **occipital margin** (Figs. 3.1d,g). **Mentum** with single or double median tooth (or a single tooth with a median notch) and 7–11 lateral teeth (Figs. 3.1b,e,h). Arrangement of mental teeth, however, gives an appearance of 3–6 median teeth. **Ventromental plates** vestigial to well developed (Figs. 3.1b,e,h). **Mandible** with one apical tooth and 4 inner teeth (Figs. 3.1c,f,i). **Premandible** apically broad with 5–7 teeth.

 | DOI:10.1163/9789004694781_004

Remarks

The most characteristic feature distinguishing subfossil *Diamesa* from other genera is the relatively large number of uniform mental teeth (Figs. 3.1a,b,e,h) and remarkably wide occipital margin (Figs. 3.1d,g).

Identification of larval *Diamesa* to morphotypes/species is very difficult even in recent larvae, where the basic separation is based on characters located on the body (e.g., presence/absence of procercus, length and number of anal setae, etc.) and follows with those on the head. Since the body is missing in the subfossils, their identification is restricted to head coloration and few structures, such as mentum and mandible, which are, due to the feeding habit of larvae (scrapers), often worn.

Morphotypes

Two main morphotypes were distinguished following Brooks et al. (2007). Furthermore, in case that mandible was present, we separated two subtypes of the *zernyi/cinerella*-type (even though most likely much more species were represented in the subfossil material).

1 Mentum with 5 (occasionally 6) median teeth, mandible with 1st and 2nd inner teeth longer than apical tooth .... *Diamesa bertrami*-type
– Mentum with 1 median tooth (may be notched medially), mandible with 2nd inner tooth never as long as apical tooth .... *Diamesa zernyi/cinerella*-type 2

2 Median mental tooth notched, slightly broader than 1st laterals; 1st inner tooth of mandible as long, or longer than apical tooth .... *Diamesa zernyi/cinerella*-type A
– Median mental tooth without a notch, subequal in size to 1st laterals; 1st inner tooth of mandible shorter than apical tooth .... *Diamesa zernyi/cinerella*-type B

***Diamesa bertrami*-type.** Head capsule entirely dark brown (Fig. 3.1a). Apex of mentum with five prominent teeth, median tooth single, second is slightly taller than first, and 8 pairs of smaller lateral teeth, last one is minute (Figs. 3.1a,b). Mandible with a distinctive shape (Fig. 3.1c): first two inner teeth exceptionally long (longer than apical tooth), second tooth is longer than first one. The *D. bertrami*-type described by Brooks et al. (2007) has apex of mentum with 6 teeth, i.e. median tooth is double. Rossaro & Lencioni (2015) report a single median tooth in some specimens of *D. bertrami*.

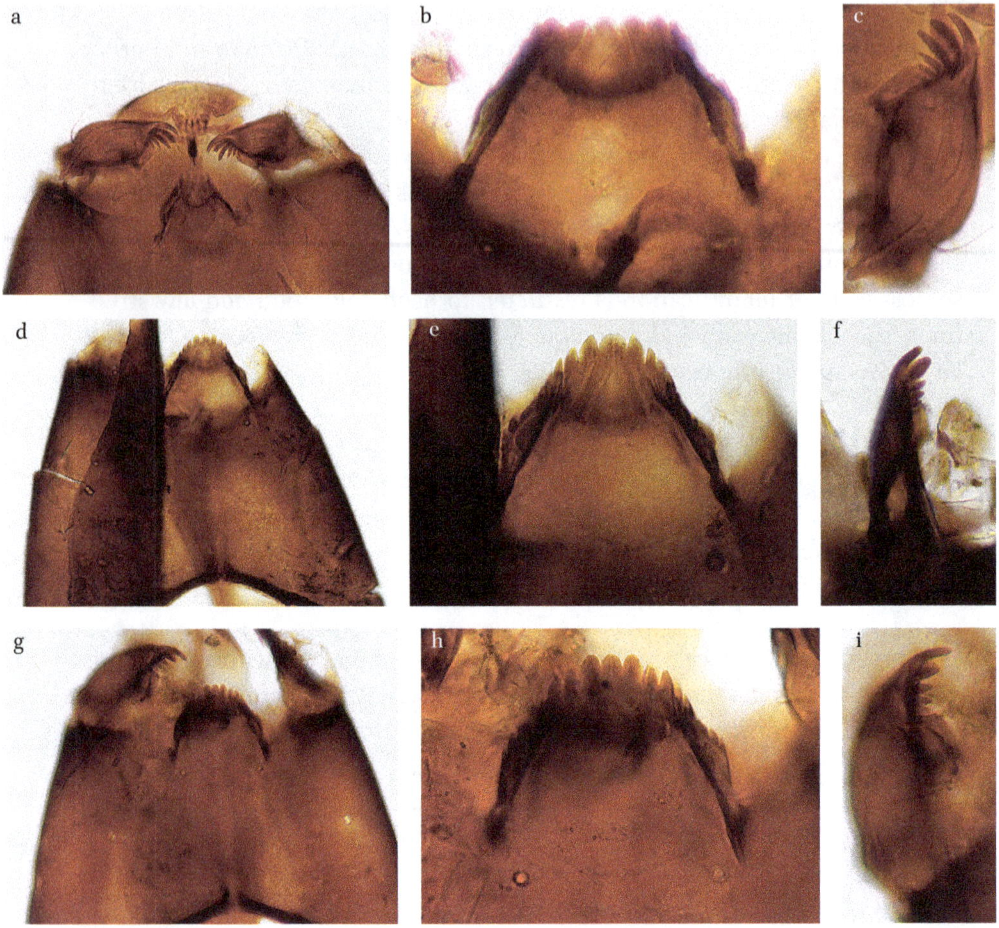

FIGURE 3.1 *Diamesa*. *Diamesa bertrami*-type: **a** head capsule with mentum, mandible and pecten epipharyngis, **b** mentum, **c** mandible. *Diamesa zernyi/cinerella*-type A: **d** head capsule, **e** mentum, **f** mandible. *Diamesa zernyi/cinerella*-type B: **g** head capsule, **h** mentum, **i** mandible

***Diamesa zernyi/cinerella*-type A.** Head capsule brown (Fig. 3.1d). Mentum with a notched median tooth and 9 pairs of lateral teeth. Median tooth is wider than first lateral, and first lateral tooth is obviously larger than second lateral. Remaining laterals are gradually decreasing (Fig. 3.1e). First inner tooth of mandible is as long, or longer (and more robust) than apical tooth (Fig. 3.1f). Ventromental plates clearly expanded laterally. In Chamutiová et al. (2020) this type is named *Diamesa* Tatra-type A.

***Diamesa zernyi/cinerella*-type B.** Head capsule dark brown (Fig. 3.1g). Mentum with a single median tooth and 9 pairs of lateral teeth. Median tooth and first pair of lateral teeth are prominent and subequal in size, other laterals are gradually decreasing (Fig. 3.1h). First inner tooth of mandible is more robust than second, but shorter than apical tooth (Fig. 3.1i). This type corresponds to *Diamesa* Tatra-type B in Chamutiová et al. (2020).

Distribution

Even though *Diamesa* remains were not abundant, they were relatively frequent, found in about one third of the surveyed Tatra Mts. lakes (SK, PL) and nearly half of the lakes in the Ukrainian part of the Eastern Carpathians regardless of the presence of an inflow stream.

Ecology and palaeoecological interpretation

Diamesa is an oligostenothermic and rheophilous genus. Larvae inhabit cold running waters, including glacial meltwater brooks; they can be found in springs and cold hygropetric sites, and in well oxygenated littoral of cold lakes (Serra-Tosio 1973, Pinder 1995 and references therein). The larvae are free-living and scrape algae and cyanobacteria from hard substrata (Berg 1995 and references therein).

The presence of *Diamesa* as cold-temperature indicators (Laroque & Hall 2003) in stratigraphic records is connected with cold climate and/or presence of inflow of cold water entering the lake (e.g., Rück et al. 1998, Gandouin & Franquet 2002, Samartin et al. 2012, Hamerlík et al. 2016).

Pseudodiamesa Goetghebuer (Fig. 3.2)

Head capsule yellow to brown with dark brown **occipital margin** (Figs. 3.2a,f). **Mentum** with single sharp median tooth and seven pairs of lateral teeth. Median tooth is separated from first laterals by a deep V-shaped notch; second lateral tooth lying dorsal to first lateral tooth seems to be partially hidden behind it. Fifth and seventh lateral teeth shorter than sixth (Figs. 3.2b,g). **Ventromental plates** joined with median and first lateral teeth extend laterally, so often only 2–4 lateral teeth are visible (Fig. 3.2b). **Mandible** with 4 inner teeth, apical tooth distinctly larger than first inner tooth (Figs. 3.2d,i). **Premandible** with 7–11 teeth (Figs. 3.2e,j).

Remarks

Three subequal teeth on the mentum apex, position of the second lateral tooth, and large ventromental plates (Figs. 3.2b,g) are distinctive for *Pseudodiamesa* among Diamesinae.

Serra-Tosio (1973) and Ilyashuk et al. (2010) distinguish two morphotypes, *P. branickii*-type and *P. nivosa*-type. Except for mentum, other diagnostic characters (antennal ratio, pecten epipharyngis, labrum, premandible, mandibular seta interna) are usually missing in subfossils, thus only the morphology of mentum can be used for the separation of morphotypes. As *Pseudodiamesa* remains are often abundant in the sediment record, the within-genus determination may be important for palaeoreconstructions due to the differences in the ecological requirements of the morphotypes.

Morphotypes

Two morphotypes were identified following Serra-Tosio (1973) and Ilyashuk et al. (2010). However, their separation in case of subfossils may be challenging or even impossible. Thus, the following key has to be used with extra caution even with 4th instar larval remains.

1 Teeth of mentum unicolored, black, 4th to 7th lateral teeth hardly visible in unflattened mentum, covered with ventromentum .... *Pseudodiamesa branickii*-type

– Mentum with a lighter median tooth compared to the black lateral teeth; outermost lateral teeth are separated and visible in unflattened mentum .... *Pseudodiamesa nivosa*-type

***Pseudodiamesa branickii*-type.** Antennal ratio (length of basal segment/length of flagellum) < 2.5 (Fig. 3.2c). Pecten epipharyngis consists of seven weakly sclerotized scales (Fig. 3.2e). Labral lamellae absent. Premandible with 7–9 teeth (Fig. 3.2e). Mandibular seta interna consists of 9–15 branches (Fig. 3.2d). All teeth of mentum unicolored, black, fourth to seventh lateral teeth are hardly visible in unflattened mentum (Fig. 3.2b).

***Pseudodiamesa nivosa*-type.** Antennal ratio ≥ 2.5 (Fig. 3.2h). Pecten epipharyngis is characterized by an even number (3–4 pairs) of strongly sclerotized scales (Fig. 3.2j). In well preserved head capsules, labral lamellae may be present between bases of SI setae (Figs. 3.2j,k). Premandible with more than 10 teeth. Mandibular seta interna consists of 14–24 branches (Fig. 3.2j). Mentum with a lighter median tooth compared to the black lateral teeth; outermost lateral teeth are more visible in unflattened mentum (Fig. 3.2g). However, it is important to note that the first and second instar larvae of both morphotypes have a broad median mental tooth that is lighter than the blackish lateral teeth (Fig. 3.2l), thus young instars are inappropriate for separating morphotypes.

Distribution

Both types were common in the lakes of the Tatra Mts. (SK, PL). In addition, *P. branickii*-type was also recorded in a lake in the Chornohora Massif (UA).

Ecology and palaeoecological interpretation

Pseudodiamesa is a cold-adapted genus. Larvae are typical inhabitants of standing and running waters in high-latitude and high-altitude regions, however they also occur in cold springs at lower altitudes.

While *P. nivosa* is associated with "glacial" conditions (extremely low temperatures, freezing and drying out of habitats), *P. branickii* is characterised by a higher temperature optimum and occurs in moderate conditions further downstream and warmer lake littorals (see Ilyashuk et al. 2010 and references therein, Hamerlík et al. 2017, Bitušík et al. 2017). In the Tatra Mts. training-set, *P. nivosa* represents the lowest (coldest) air temperature optimum (July average 7.5 °C), while the T-optimum of *P. branickii* is considerably higher, 9.2 °C (unpublished data).

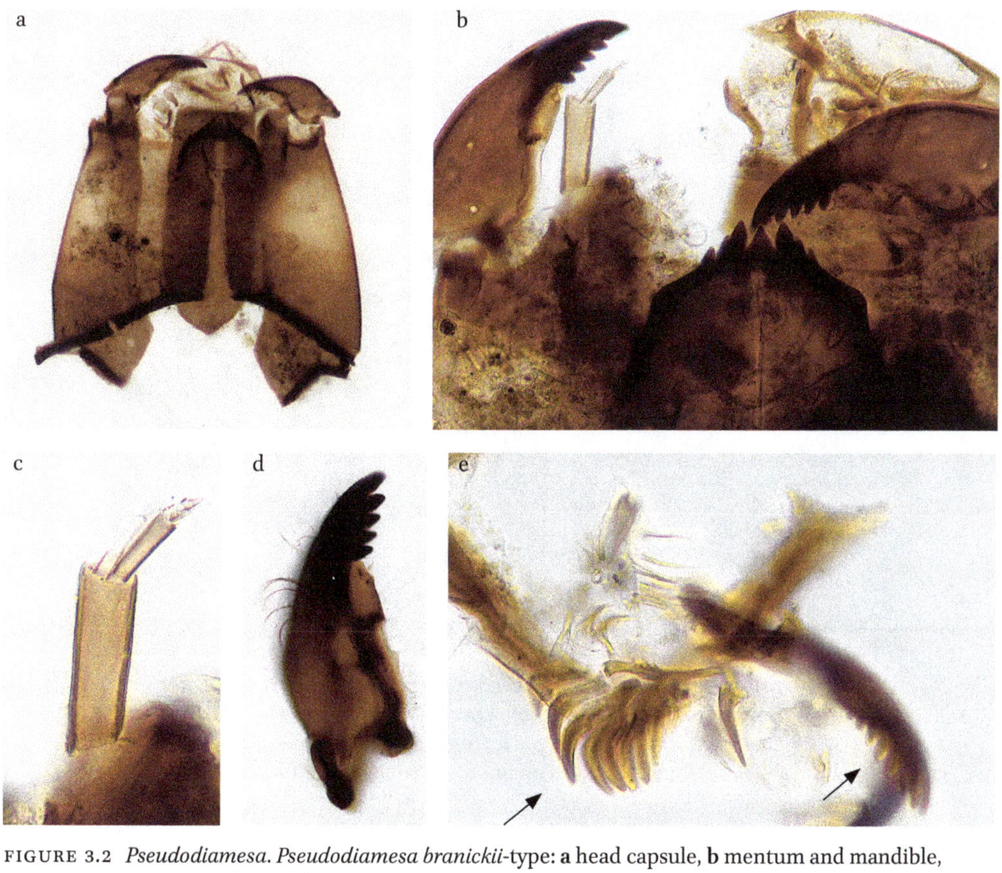

FIGURE 3.2 *Pseudodiamesa. Pseudodiamesa branickii*-type: **a** head capsule, **b** mentum and mandible, **c** antenna, **d** mandible, **e** detail of pecten epipharyngis and premandible (arrows indicate seven scales of pecten epipharyngis and 8–9 teeth of the premandible)

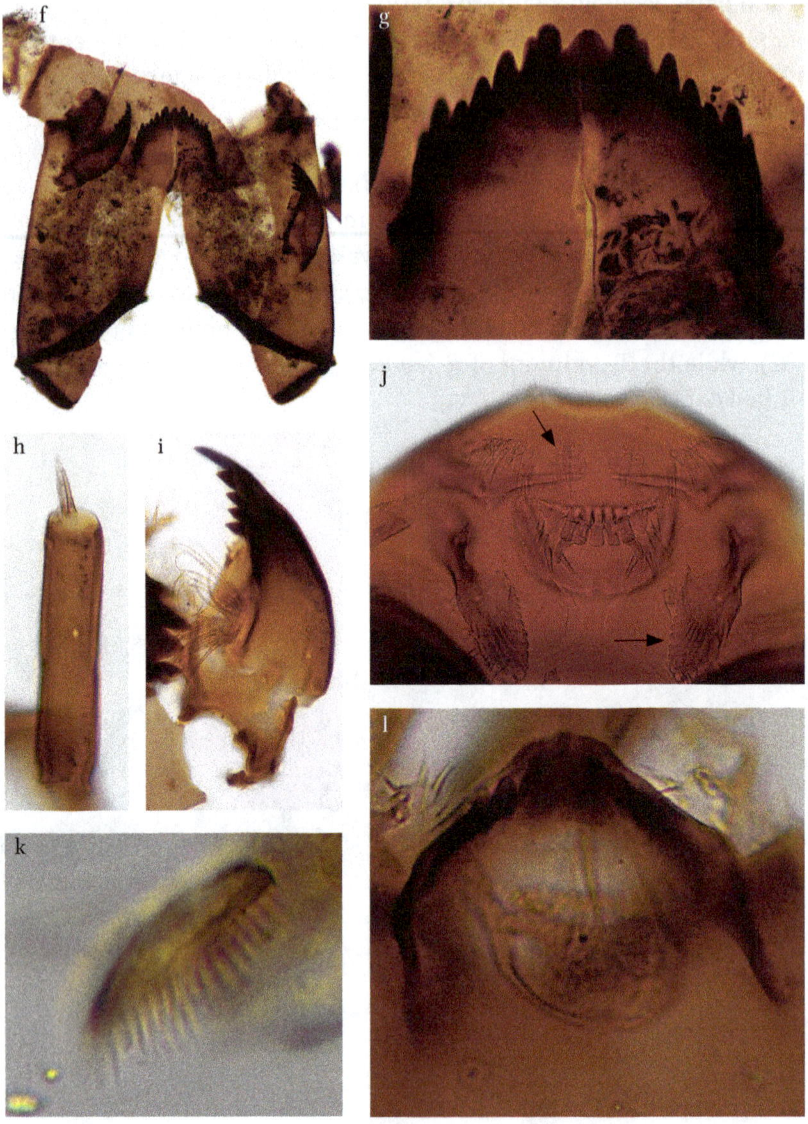

FIGURE 3.2 (continuation) *Pseudodiamesa nivosa*-type: **f** head capsule, **g** mentum, **h** antenna, **i** mandible, **j** detail of pecten epipharyngis and premandible (arrows indicate premandible teeth, SI setae and labral lamellae), **k** detail of labral lamella, **l** mentum of a young instar *Pseudodiamesa*

Pseudodiamesa remains often represent an important part of the subfossil chironomid assemblage, especially in sediments corresponding to the cold phases of the Late Glacial and Holocene periods (e.g., Lotter et al. 2000, 2006, Filippi et al. 2007, Bitušík et al. 2009 and unpublished data).

Pseudokiefferiella parva (Edwards, 1932) (Fig. 3.3)
Head capsule dark brown (Fig. 3.3a). **Mentum** striped, with one median tooth and 7 pairs of lateral teeth (outermost tooth may be hardly visible). Median tooth is subequal to first lateral tooth in size, second lateral tooth is obviously smaller and fused to the first lateral. There is a distinct space between the lighter second and darker third laterals; third to seventh laterals are darker and create a separate cluster, outermost lateral tooth minute, not always visible (Fig. 3.3b). **Ventromental plates** are small. **Setae submenti** situated below bases of mentum. First inner tooth of **mandible** is as long as apical tooth (Fig. 3.3b).

Remarks

Larvae of *Pseudokiefferiella* are similar to *Diamesa* and the only reliable diagnosis is based on larval body. Comparing subfossil heads with recent larval material, we suggest that remains of both genera are distinguishable by the shape and coloration of mentum but this character should be used with caution.

Distribution

Small number of subfossils remains were recorded in several very cold Tatra Mts. lakes (SK, PL).

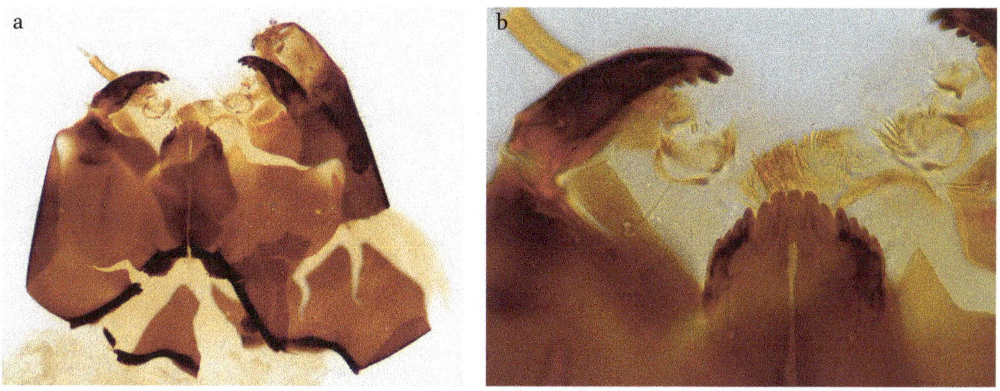

FIGURE 3.3 *Pseudokiefferiella*. **a** head capsule with mentum, mandible and antenna, **b** mentum and mandible

Ecology and palaeoecological interpretation

The single Holarctic species described is distinctly cold-stenothermic and rheophilic. Larvae inhabit glacial brooks (Lencioni & Rossaro 2003), cold alpine streams and lake inlets and outlets (Novikmec et al. 2015); they are also reported as colonisers of mosses in spring-fed streams (Kownacki 1987). Together with *P. nivosa,* it belons to the most cold-stenothermic morphotypes in the Tatra Mts. training-set (mean July air temperature 7.6 °C, unpublished data).

In the stratigraphic records, *Pseudokiefferiella* belongs to cold-temperature indicators, and can be indicative of a significant impact of cold water entering the lake. Larvae may live in lake littorals near the inflow or may be transported into fossil assemblages by tributaries.

CHAPTER 4

Prodiamesinae

Head capsule of the subfamily Prodiamesinae differ from that of Chironominae, Orthocladiinae and Diamesinae by large ventromental plates without striae and with numerous and dense setae creating a ventromental beard, often visible even under low magnification (Fig. 4.1).

Prodiamesa Kieffer (Fig. 4.1)
Head capsule large, yellow with black **occipital margin** and with dark brown pattern on postmentum (Fig. 4.1a). **Mentum** consisting of 2 low median teeth and 8 pairs of lateral teeth; first lateral tooth tall, trifid, 6th lateral higher than 5th and 7th, outermost lateral minute (Fig. 4.1b). **Ventromental plates** long and broad with long beard exceeding the edge of the head, well visible under lower magnification (Figs. 4.1a,b). **Mandible** with one apical tooth and 4 inner teeth (Fig. 4.1b).

Remarks
The unique shape of mentum together with the well-developed ventromental plates bearing dense and long beard distinguish this morphotype from other chironomids.

Morphotypes
One morphotype was identified.
***Prodiamesa olivacea*-type.** We propose to establish this morphotype by assuming that it could consist of *P. olivacea* (Meigen, 1818) and its sibling species *P. bureschi* Michailova, 1977 and can be distinguishable from *P. rufovittata* Goetghebuer, 1932 by the dense setation (beard) of the mentum (Figs. 4.1a,b).

Distribution
Small number of subfossil remains were collected from the reservoirs in the Štiavnické vrchy Mts. (SK), several alpine lakes in the Tatra Mts. (SK, PL) and in one lake in the Chornohora Massif (UA).

 | DOI:10.1163/9789004694781_005

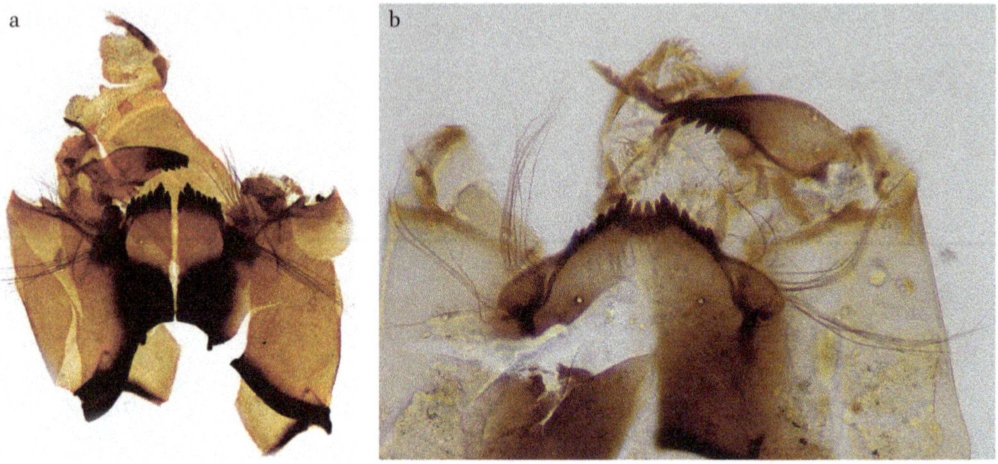

FIGURE 4.1 *Prodiamesa olivacea*-type: **a** head capsule, **b** detail of head capsule showing mentum, mandible and ventromental plates with characteristic long beard

Ecology and palaeoecological interpretation

Generally, *Prodiamesa* is a rather ubiquitous genus occurring in a wide range of water bodies from springs through streams and rivers to stagnant waters of various sizes, i.e., pools, ponds and lakes.

P. olivacea (Meigen, 1818) is obviously the most common species recorded in both oligotrophic alpine lakes and mesotrophic to eutrophic lakes and reservoirs at lower altitude (Langdon et al. 2006, Chamutiová et al. 2018), even at polluted sites in rivers (Ilkova et al. 2016).

Prodiamesa subfossils are found relatively rarely in the subfossil material. Its occurrence in the sediment sequence may signalize the presence of an inflow stream.

CHAPTER 5

Orthocladiinae

Head minute to large, yellow to brown or brown-black, occipital margin can be darkened, dark brown or black. Antenna not longer than head with 4–7 (usually 5) segments, third segment never annulate. Labrum and pecten epipharyngis with variably shaped structures (labral setae, labral lamellae) important for identification, however these characteristics usually missing from subfossils. Premandible with 1–6 teeth. Mandible with single apical tooth and usually 3 inner teeth. Mentum is very variable, mostly curved, with 3–29 teeth; mainly 4–6 pairs of lateral teeth are present. Ventromental plates vestigial to relatively large, never with striae; setal beard is present in some genera.

Orthocladiinae head capsules can be separated from Chironominae by the absence of striated ventromental plates, and by the eyespots, which are usually

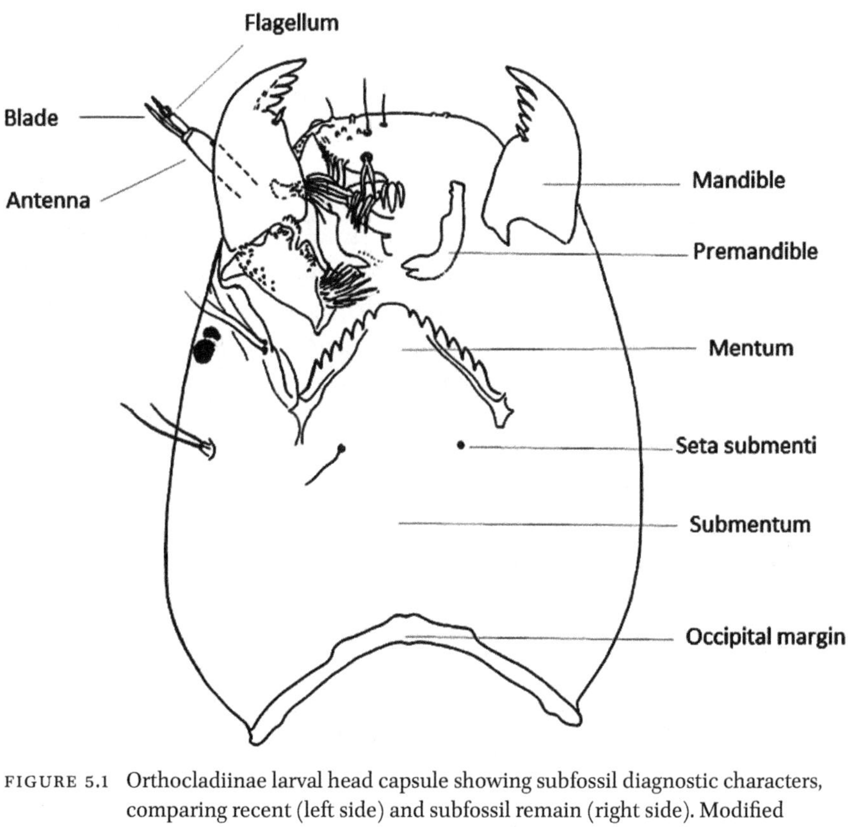

FIGURE 5.1 Orthocladiinae larval head capsule showing subfossil diagnostic characters, comparing recent (left side) and subfossil remain (right side). Modified after Bitušík (2000)

 | DOI:10.1163/9789004694781_006

contiguous or, when separate, with a dorsal eye-spot posterior to a ventral eye-spot. Diamesinae can be distinguished from orthoclads by having annulate third antennal segment (unfortunately, antennas are usually missing in subfossils), generally higher number of lateral teeth and more robust head occipital margin. The most important characters for separation of subfossil orthoclads are the shape, number and arrangement of mental teeth together with the shape of ventromental plates.

Most of recorded head capsules belong to rheophilic to rheobiontic genera; lentic and semiterrestrial/terrestrial genera are present to a less extent.

KEY TO GENERA OF SUBFOSSIL ORTHOCLADIINAE

1 Mentum with single median tooth 2
– Mentum with 2 or more median teeth 9

2 Mentum with 2 pairs of lateral teeth and an exceptionally large triangular median tooth .... *Orthocladius* (*Symposiocladius*) *lignicola*
– Mentum with more than 2 pairs of later leeth 3

3 Mentum with 4 pairs of lateral teeth 4
– Mentum with 5 and more pairs of lateral teeth 5

4 Mentum arched, usually with clearly visible stripes, median tooth 2–3 times as broad as 1st lateral tooth, ventromental plates long and straight .... *Eukiefferiella* in part (*E. devonica*-type)
– Mentum rather horizontal, without stripes, median tooth broader, about 4 times as broad as 1st lateral tooth, ventromental plates short and curved laterally .... *Pseudosmittia*

5 Mentum with 5 pairs of lateral teeth 6
– Mentum with 6 and more pairs of lateral teeth 8

6 Weakly pigmented ventromental plates extending lateral to outer teeth, broadly triangular at apex, sparse beard may be present; median mental tooth broad with indentations on apical margin .... *Psectrocladius* in part (*P. flavus*-type)
– Ventromental plates weak or with a dark triangular knob below outer lateral teeth, median mental tooth broad or narrow, without indentations on apical margin 7

7 Mentum strongly arched, with stripes below teeth, lateral teeth subequal in size .... *Eukiefferiella* in part (*E. brevicalcar*-type, *E. fittkaui*-type)
– Mentum rather horizontal, without obvious stripes below teeth, 1st lateral tooth clearly larger than remaining laterals .... *Smittia/Parasmittia*

8 Ventromental plates obviously swollen and rounded at apex .... *Parakiefferiella*
– Ventromental plates narrow, elongate; median tooth broad, low, dome-shaped and paler than remaining teeth (beard may be present) .... *Paracladius*
– Ventromental plates narrow, not broadening towards apex; median tooth of mentum not dome-shaped (beard never present) .... *Cricotopus/Orthocladius*

9 Mentum with 2 median teeth 10
– Mentum with 3 or more median teeth 27

10 Mentum with 4 pairs of lateral teeth 11
– Mentum with 5 or more pairs of lateral teeth 13

11 Mentum of prolonged triangular shape, median mental teeth prominent, about as long as the total lengths of lateral teeth, ventromental plates long and narrow .... *Synorthocladius*
– Mentum less arched, median mental teeth only slightly longer than 1st lateral tooth 12

12 Ventromental plates enlarged, extending beyond lateral margin of mentum .... *Gymnometriocnemus/Bryophaenocladius*
– Ventromental plates weak, not extending beyond mentum laterally .... *Paratrissocladius*

13 Mentum with 5 pairs of lateral teeth 14
– Mentum with 6 pairs of lateral teeth (1st lateral may be minute) 24

14 Ventromental plates narrow, not expanding towards apex 15
– Ventromental plates large, expanding towards apex (bulbous, truncate, foot-shaped or triangular) 19

15 Mentum with 4th and 5th lateral teeth appressed to each other (may look like 4 lateral teeth with the outermost apically notched), distinct dark oval patch on submentum of well sclerotized specimens .... *Brillia*

– Mentum with 4th and 5th lateral teeth clearly separated 16

16 Mentum with bulbous, darkly pigmented tubercle situated on either side of the base .... *Limnophyes*/*Paralimnophyes*
– No such tubercle at the base of mentum 17

17 Median teeth of mentum shorter than 1st lateral tooth, mentum not striated .... *Metriocnemus*
– Median teeth of mentum longer than 1st lateral tooth, mentum striated 18

18 Head capsule darkly pigmented, seta submenti located close to bases of mentum .... *Eukiefferiella* in part (*E. claripennis*-type T)
– Head capsule light, seta submenti located far below bases of mentum .... *Tvetenia*

19 Ventromental plates exceptionally long, about as long as the length of mentum .... *Nanocladius*
– Ventromental plates shorter, if extending below mentum, than by less than half the length of mentum 20

20 Ventromental plates bulbous at apex 21
– Ventromental plates of different shape at apex (triangular, truncate or foot-shaped) 22

21 Mentum strongly arched, submentum of the same color as the rest of head capsule, outermost lateral tooth reduced in size .... *Chaetocladius*
– Mentum rather triangular, submentum darker than the remainder of the head capsule (may not be clear on weakly sclerotized individuals), outermost lateral teeth not reduced significantly .... *Heterotrissocladius*

22 Ventromental plates foot-shaped, outer margin bent upwards, 3rd and 4th lateral teeth subequal .... *Parametriocnemus*/*Paraphaenocladius*
– Ventromental plates triangular or truncate, lateral teeth decreasing gradually, 3rd tooth taller than 4th 23

23 Ventromental plates triangular at apex, beard, if present, short .... *Psectrocladius* in part (*P. psilopterus*-type, *P. sordidellus*-type)
– Ventromental plates truncate at apex, with long dense beard .... *Rheocricotopus* in part (*R. effusus*-type)

24 Mentum with 1st lateral tooth reduced 25
– Mentum with 1st lateral tooth normally developed 26

25 1st lateral mental tooth shorter than 2nd tooth .... *Zalutschia*
– 1st lateral mental tooth taller than 2nd tooth .... *Rheocricotopus* in part (*R. fuscipes*-type)

26 Mentum horizontal, 2nd lateral mental tooth reduced, shorter than 1st and 3rd laterals, ventromental plates with dense beard .... *Diplocladius*
– Mentum arched, lateral mental teeth decreasing gradually, 2nd lateral tooth taller than 3rd lateral tooth, ventromental plates without beard .... *Hydrobaenus*

27 Mentum with 3 median teeth 28
– Mentum with 4 or 6 median teeth 33

28 Mentum with 4 pairs of lateral teeth, ventromental plates narrow and long, about half the length of mentum .... *Parorthocladius*
– Mentum with 5 pairs of lateral teeth 29

29 Mentum with 1st lateral teeth obviously reduced .... *Corynoneura* in part (*C. lobata*-type)
– Mentum with 1st lateral teeth not reduced 30

30 Central tooth of mentum strongly reduced 31
– Central median tooth subequal to outer pair of median teeth or only slightly shorter 32

31 Head capsule without reticulation, 4th and 5th lateral teeth of mentum appressed to each other (may appear like 4 lateral teeth with the outermost apically notched), distinct dark oval patch on submentum of well sclerotized specimens .... *Brillia*
– Head capsule with net-like reticulation, 4th and 5th lateral teeth of mentum well separated, not appressed to each other, no dark patch on submentum .... *Corynoneura* in part

32 Head capsule light, central median tooth slightly shorter than outer median teeth .... *Corynoneura* in part
– Head capsule darker, central median tooth subequal to outer pair of median teeth .... *Thienemanniella*

33 Mentum with 4 median teeth, 1st lateral teeth reduced in size .... *Diplocladius*
– Mentum with 6 median teeth .... *Epoicocladius*

Note that some genera with unclear number of median mental teeth are keyed twice (e.g., *Brillia* with both 2 and 3 medial teeth and *Diplocladius* with 2 and 4 median teeth).

Brillia Kieffer (Fig. 5.2)
Head capsule triangular and yellow (Fig. 5.2a). **Antenna** with 4 segments, second segment divided into a small proximal and a longer (~ 3×) distal part (Fig. 5.2c). **Premandible** bifid, weak notch between teeth. Mandible with 4 inner teeth. **Mentum** dark brown with two long median teeth, a very small tooth usually present in between bases of them, and 5 pairs of lateral teeth (Fig. 5.2b). **Ventromental plates** small, triangular. **Setae submenti** positioned on the submentum closer to occipital margin than base of mentum (Fig. 5.2a).

Remarks
Larval *Brillia* remind *Euryhapsis* and *Eurycnemus* (neither of them shown in this book) and can be distinguished from them by details on antennae and labrum, though these structures are usually absent in subfossil material. Owing to their ecology (both are rheobionts and *Eurycnemus* are obligate ectoparasites (Andersen et al. 2013)), the occurrence of *Euryhapsis* and *Eurycnemus* in lake sediments is unlikely.

Two subfossil morphotypes of *Brillia* can be distinguished corresponding to *B. modesta* and *B. flavifrons* species groups following Oliver and Roussel (1983). As the valid name of *B. modesta* is currently *B. bifida*, the recorded morphotype is named *B. bifida*-type.

Morphotypes
One morphotype was recorded.
***Brillia bifida*-type**. Head capsule yellow, occipital margin blackish. Well sclerotized specimens have distinct brown-black oval patch on the postmentum (Fig. 5.2a). Fourth and fifth lateral teeth of mentum are adpressed to each other (Fig. 5.2b).

Distribution
The morphotype was rare, recorded only in few Tatra Mts. lakes (SK) and in two reservoirs in the Štiavnické vrchy Mts. (SK).

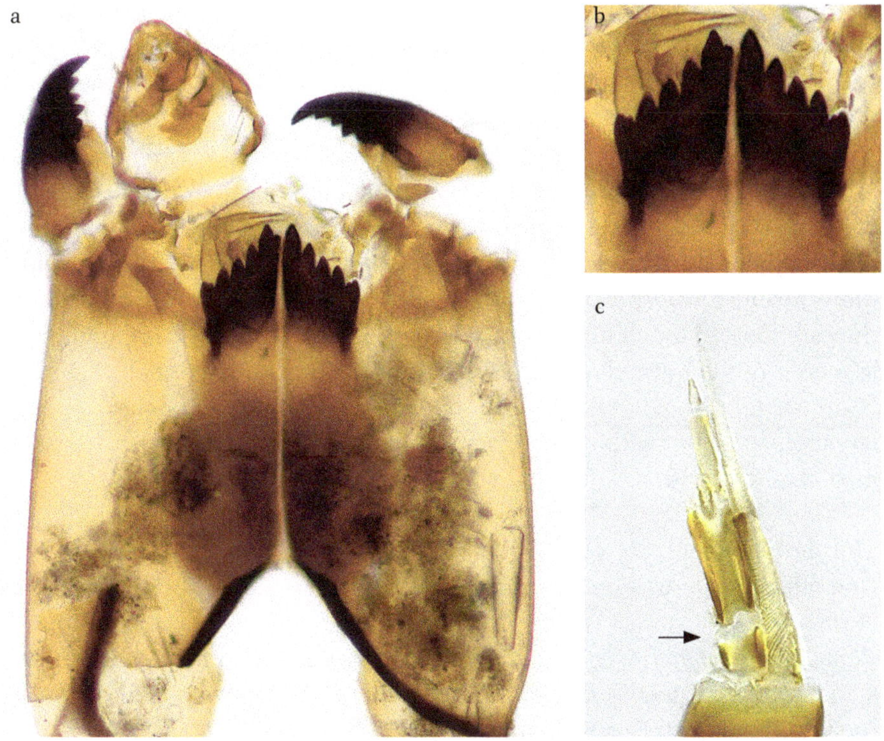

FIGURE 5.2 *Brillia bifida*-type. **a** head capsule (arrow indicates position of seta submenti), **b** mentum, **c** flagellum of antenna with subdivided 2nd antennal segment (indicated by arrow)

Ecology and palaeoecological interpretation

Larval *Brillia* live in the littoral of lakes and in different kinds of running waters with pH ranging from neutral to acidic (pH 4–7). Larvae are predominantly shredders (Janecek & Moog 2002) and are associated with submerged wood and decaying leaves.

Brillia remains in subfossil samples will indicate a shaded inlet stream with high oxygen content, and/or a littoral zone with slowly decaying leaves and wood (Moller Pillot 2013).

Chaetocladius Kieffer (Fig. 5.3)

Head yellow to light brown with dark **occipital margin** (Fig. 5.3a). **Antenna** 5-segmented, never strongly reduced. **Premandible** bifid with 1–2 blunt inner teeth (Fig. 5.3f). **Labral lamellae** of various shapes situated between labral setae SI, usually well-developed but occasionally vestigial or absent (Fig. 5.3c). Apical **mandibular** tooth shorter than combined width of 3 inner teeth (Fig. 5.3d). **Mentum** with 1–2 median teeth and 5 pairs of lateral teeth, outermost tooth may be reduced (Figs. 5.3b,e). **Ventromental plates** usually large, rounded and extending beyond outer lateral mental tooth (Fig. 5.3e).

Remarks

Heterotrissocladius (Fig. 5.10), *Parametriocnemus* and *Paraphaenocladius* (Fig. 5.18) also have mentum with double median tooth, and 5 pairs of laterals, combined with large ventromental plates extending beyond the outer lateral teeth without beard. Setae submenti of *Chaetocladius* are positioned under the ventromental plates, while in *Parametriocnemus* and *Paraphaenocladius* they are located much higher (for comparison see Fig. 5.18b). The outermost mental tooth is not reduced in *Heterotrissocladius*. *Metriocnemus* (Fig. 5.13), *Hydrobaenus* (Fig. 5.11) and *Thienemannia* (not shown in this book) have similar menta, however without large ventromental plates and obviously narrower median teeth/tooth.

Morphotypes

One morphotype was found in the subfossil material following Brooks et al. (2007).

***Chaetocladius piger*-type** (Fig. 5.3a). Mentum with paired median tooth, and 5 pairs of lateral teeth, outermost lateral is minute. Ventromental plates are broad and bulbous (Figs. 5.3b,e).

Distribution

The morphotype was common in the Tatra Mts. (SK, PL) and also occurred in a reservoir in the Štiavnické vrchy Mts. (SK).

Ecology and palaeoecological interpretation

Most species of *Chaetocladius* are considered semiaquatic, living in seepages and hygropetric habitats; larvae of the few truly aquatic species are known predominantly from springs, brooks and streams (Janecek & Moog 2002, Bitušík & Hamerlík 2003, Moller Pillot 2013). Larvae were also recorded in the littoral of Tatra Mts. lakes and ponds in higher elevations with significant fluctuation of water level (Hamerlík et al. 2017).

Its presence in the subfossil material may indicate strong inlet influence and water level fluctuations.

Corynoneura Winertz, 1846 (Fig. 5.4)

Head capsule very small (thus easy to overlook), narrow and yellow, often with reticulation on its surface (Fig. 5.4d). **Occipital margin** pale to black (Figs. 5.4a,e,h). **Antennae** 4-segmented, subequal or longer than head, second and third segments are frequently darkened. **Premandible** with up to 12 minute teeth. **Mandible** with apical tooth smaller than any of 4 inner teeth

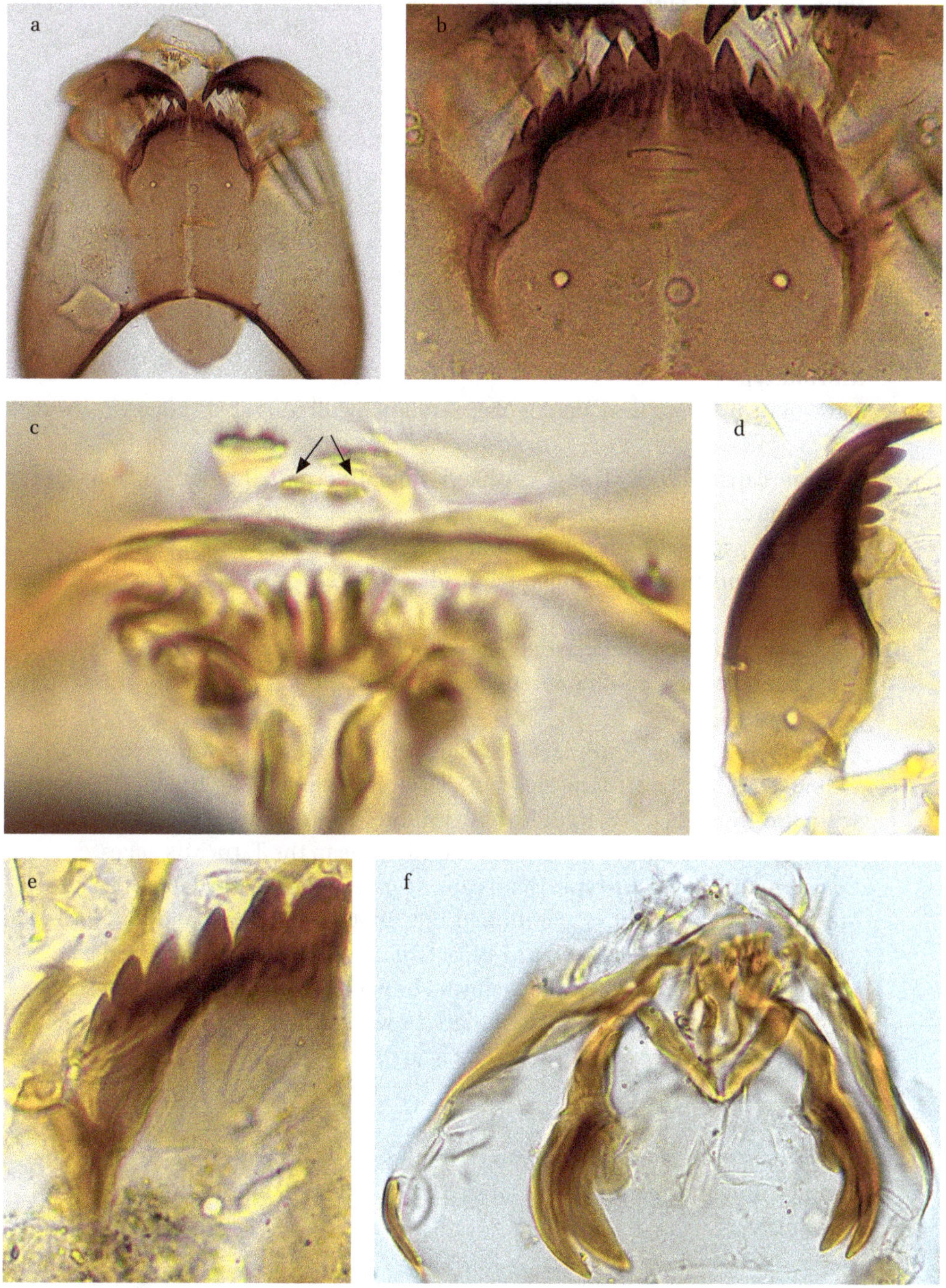

FIGURE 5.3 *Chaetocladius piger*-type: **a** head capsule, **b** mentum, **c** labral lamellae (indicated by arrows), **d** mandible, **e** detail of mentum showing minute 5th lateral tooth and bulbous ventromental plate, **f** bifid premandibles with 2 inner teeth

(Figs. 5.4c,g,j). **Mentum** with 2–3 median teeth and 5 pairs of lateral teeth (Figs. 5.4b,f,i). **Ventromental plates** narrow.

Remarks

The unique feature of the genus is the extremely long 4-segmented antenna (often broken off in the subfossil material). *Thienemanniella* (Fig. 5.26) has similar shape of mentum, but the 5-segmented antenna is shorter than head, moreover, it is without reticulation on the head capsule. It may not be a rule, but according to our experience, *Thienemanniella* head capsules are in general darker, while those of *Corynoneura* less sclerotized, paler.

Morphotypes

Three morphotypes were distinguished following Brooks et al. (2007).

1 Head capsule with strong, net-like reticulation, middle tooth of mentum minute .... *Corynoneura arctica*-type
– Head capsule without reticulation (or only weakly indicated), middle tooth of mentum reduced or subequal to adjacent teeth 2

2 1st lateral mental tooth noticeably reduced .... *Corynoneura lobata*-type
– 1st lateral mental tooth not reduced .... *Corynoneura edwardsi*-type

***Corynoneura arctica*-type.** Head capsule yellow (Fig. 5.4a), with strong, clearly visible net-like reticulation (Fig. 5.4d). Occipital margin dark brown. Mentum with three median teeth, middle tooth minute, 5 pairs of lateral teeth subequal (Fig. 5.4b). This is the most frequent morphotype in the Tatra Mts. lakes.
***Corynoneura edwardsi*-type.** Head capsule pale and not reticulated (Fig. 5.4e). Mentum similar to above mentioned morphotype, however middle tooth is more developed than in *C. arctica*-type. Lateral teeth subequal (Fig. 5.4f).
***Corynoneura lobata*-type.** Head capsule brownish with weak wrinkled sculpturing (Fig. 5.4h). Occipital margin dark brown. Middle tooth and first lateral tooth of mentum reduced in size (Figs. 5.4i,j).

Distribution

Corynoneura remains were common in all lake districts (SK, PL, UA), absent perhaps only in the uppermost lakes of the Tatra Mts. and in the Maramures Mts. (RO).

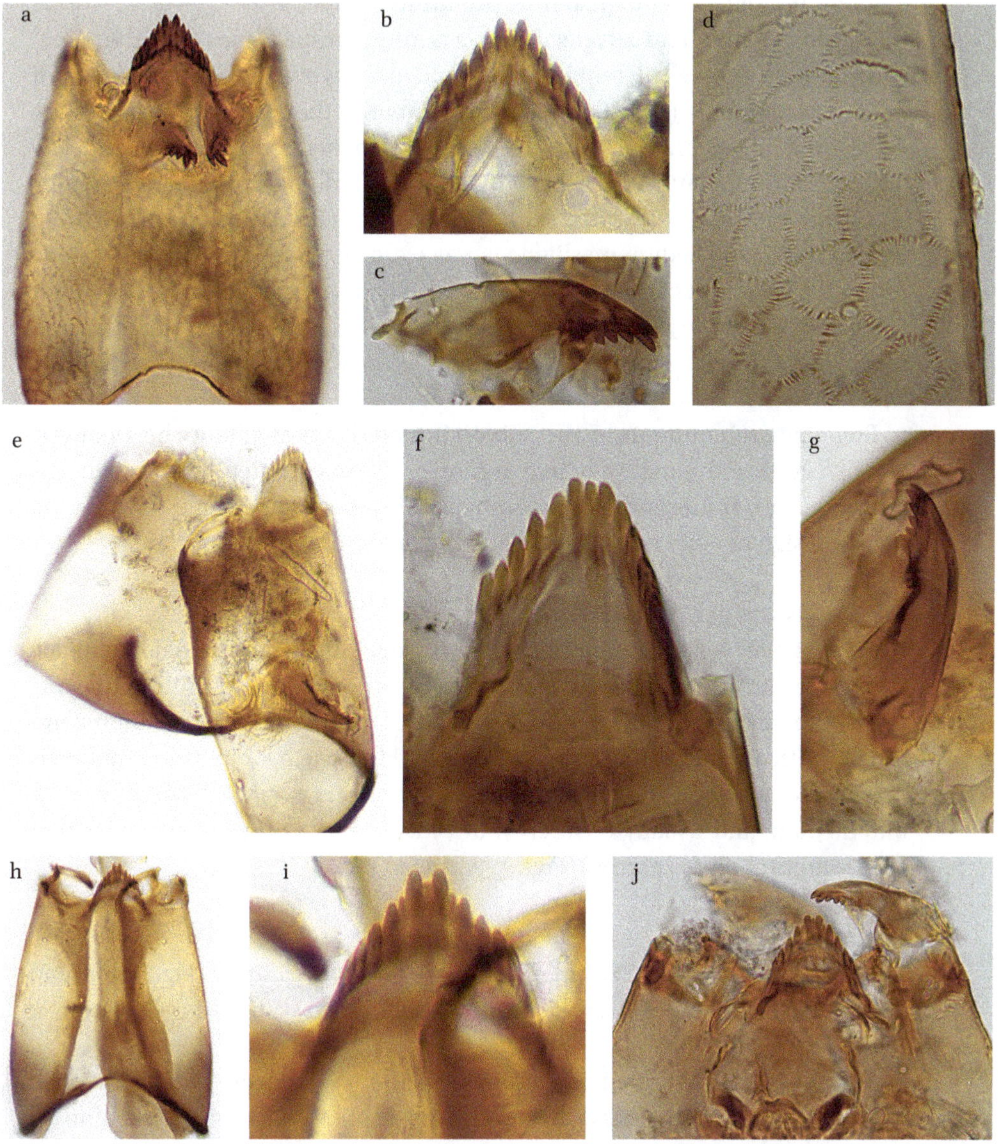

FIGURE 5.4 *Corynoneura. Corynoneura arctica*-type: **a** head capsule, **b** mentum, **c** mandible, **d** detail of head capsule showing net-like reticulation. *Corynoneura edwardsi*-type: **e** head capsule, **f** mentum, **g** mandible. *Corynoneura lobata*-type: **h** head capsule, **i** mentum, **j** mentum and mandible

Ecology and palaeoecological interpretation

It is a genus with about 20 species known from Europe (Spies & Sæther 2013) with various ecological requirements and occupying virtually all types of aquatic ecosystems (Bitušík 2000 and references therein). Lotic species are considered oligostenothermic, and sensitive to organic pollution and low oxygen content. Lentic species live in the littoral of oligotrophic to eutrophic lakes and are often associated with macrophytes (Brooks et al. 2007, Moller Pillot 2013) or soft sediments; living larvae were also found in deeper layers of profundal sediments of some Tatra Mts. lakes (unpublished data). Due to their small size, larval *Corynoneura* are polyvoltine and appear to have high dispersal capacity (Moller Pillot 2013) making them ready colonizers of temporary waters.

Remains are common in the subfossil material throughout the Holarctic (Brooks et al. 2007) and they may indicate well oxygenated littoral with soft sediment and/or presence of macrophytes. Mass occurrence for a short period may be a sign of early colonization after an environmental disturbance.

Cricotopus van der Wulp (Fig. 5.5)

Head capsule yellow to blackish brown, **occipital margin** pale to black (Figs. 5.5a,c,f,i,l,o,s). **Antenna** usually 5-segmented. **Premandible** with one, rarely two apical teeth. **Mandible** with short apical tooth (shorter than combined width of inner teeth) and 3 inner teeth (Figs. 5.5e,h,k,n,r). Outer margin is usually crenulated (Fig. 5.5e), sometimes smooth (Figs. 5.5h,n). **Mentum** with one median, and usually 6 (rarely 5 or 7) lateral teeth (Figs. 5.5b,d,g,j,m,p,t). **Ventromental plates** narrow (Figs. 5.5b,d,g,j,m,p,t).

Remarks

Cricotopus larvae are extremely similar to *Orthocladius* (Fig. 5.15) and *Paratrichocladius* (recently placed in *Cricotopus* as subgenus, Cranston & Krosch 2015).

Tuft of setae on the body segments of most species can distinguish *Cricotopus* from *Orthocladius*, however this feature is inapplicable in subfossil material, and the same goes for characters on antennae (Andersen et al. 2013b). In general, the following characters seem to be more typical for *Cricotopus* than *Orthocladius* (though can be present in some *Orthocladius* as well): outer margin of mandible strongly crenulated, second lateral tooth smaller than first one and/or reduced and fused to first lateral tooth, and outer four lateral teeth forming a distinct group. Brooks et al. (2007) state that lateral teeth of *Cricotopus* are usually more rounded and pointed apically and the median tooth is relatively narrow, while in general, *Orthocladius* have lateral teeth more rectangular and median tooth can be broad (> 3 times broader than first lateral). *Paratrichocladius* is usually distinguished by having the first lateral tooth of the mentum constricted at base so that broader in the middle than at the base, while normally widest at the base in *Cricotopus* and *Orthocladius*.

Cuppen and Tempelman (2018) proposed a key to distinguish *Cricotopus* subgenera using characters that can also be applied to sub-fossil remains. Subgenus *Isocladius*: median mental tooth usually rounded, triangular in shape, always < 2.5 times as wide as first lateral tooth; second lateral mental tooth strongly reduced and for the largest part fused with the first lateral tooth. Subgenus *Cricotopus*: median mental tooth of varying size, either small or up to 3 or 4 times as wide as first lateral tooth, second lateral tooth not reduced (e.g., Figure 39).

Morphotypes

Seven morphotypes were recognized.

1 2nd lateral tooth not significantly smaller than adjacent teeth, outermost 4 lateral teeth not forming a distinct cluster; mentum dark, uniformly colored; median tooth narrow (about as tall as wide) only slightly broader than 1st lateral tooth .... *Cricotopus tremulus*-type

– 2nd lateral tooth significantly smaller than adjacent teeth and outermost 4 laterals forming a distinct cluster; mentum uniformly colored or median part lighter than darker outermost 4 lateral teeth 2

2 2nd lateral tooth of mentum reduced and appressed to 1st lateral tooth 3

– 2nd lateral tooth of mentum not reduced (but smaller than adjacent teeth) and clearly separate from 1st lateral teeth 5

3 Median tooth of mentum triangular, about same height as 1st lateral tooth, 1st and 3rd lateral teeth subequal .... *Cricotopus obnixus*-type

– Median tooth of mentum rounded, taller than 1st lateral tooth, 3rd lateral tooth shorter than 1st lateral 4

4 Median mental tooth markedly taller than 1st lateral tooth .... *Cricotopus sylvestris*-type

– Median mental tooth only slightly taller than 1st lateral tooth .... *Cricotopus intersectus*-type

5 Mentum strongly arched with median and first two lateral teeth of mentum lighter than dark outermost 4 laterals, median tooth broad (3–4 × width of 1st lateral tooth) and strongly angled at mid-height (along with 1st lateral tooth) .... *Cricotopus cylindraceus*-type
– Mentum rather horizontal, uniformly colored, median tooth up to 2× as wide as 1st lateral tooth, circular, 1st laterals "pear-like", broader mid-height than at base 6

6 Median mental tooth broader, about 2× as broad as tall, seta submenti between bases of ventromental plates .... *Cricotopus skirwithensis*-type
– Median mental tooth narrower, not more than 1.5× broader than 1st lateral tooth, seta submenti below bases of ventromental plates .... *Cricotopus* type Tatra

***Cricotopus* (*Cricotopus*) *cylindraceus*-type** (Fig. 5.5a). Mentum strongly arched with median and first 2 lateral teeth of mentum lighter than dark outermost 4 laterals, median tooth broad (3–4 × width of first lateral tooth) and triangular, strongly angled at mid-height (Fig. 5.5b).
***Cricotopus* (*Cricotopus*) *tremulus*-type.** Head capsule dark brown with black occipital margin and mandibular teeth (Fig. 5.5c). Teeth of mentum dark brown to black in color; single narrow median tooth, equal to or only slightly wider than the first lateral tooth (Fig. 5.5d). Mandible with crenulated outer margin (Fig. 5.5e). Premandible with single tooth.
***Cricotopus* (*Isocladius*) *intersectus*-type.** Head capsule yellow with dark brown occipital margin (Fig. 5.5f). Mentum single broad median tooth taller than the first lateral tooth. Second lateral tooth minute and fused with first lateral tooth. Median tooth and first two laterals form a distinct group separated from the other laterals (Fig. 5.5g).
***Cricotopus* (*Isocladius*) *obnixus*-type** (Fig. 5.5i). Mentum characteristic, broad and dark; apex of mentum roughly horizontal as median tooth, first and third laterals lie nearly in line; second lateral tooth very short while third tooth robust, wider than outer laterals (Fig. 5.5j). Mandible markedly hooked at apex (Fig. 5.5k).
***Cricotopus* (*Isocladius*) *sylvestris*-type** (Fig. 5.5l). Head capsule with very large rounded median tooth and 6 pairs of lateral teeth. Second lateral tooth is small and fused with the first one (Fig. 5.5m). Premandible apically bifid.
***Cricotopus* (*Paratrichocladius*) *skirwithensis*-type.** Head capsule light brown with dark brown occipital margin and mandibular teeth (Fig. 5.5o). Mentum triangular with a single median tooth up to 2 times width of first lateral tooth.

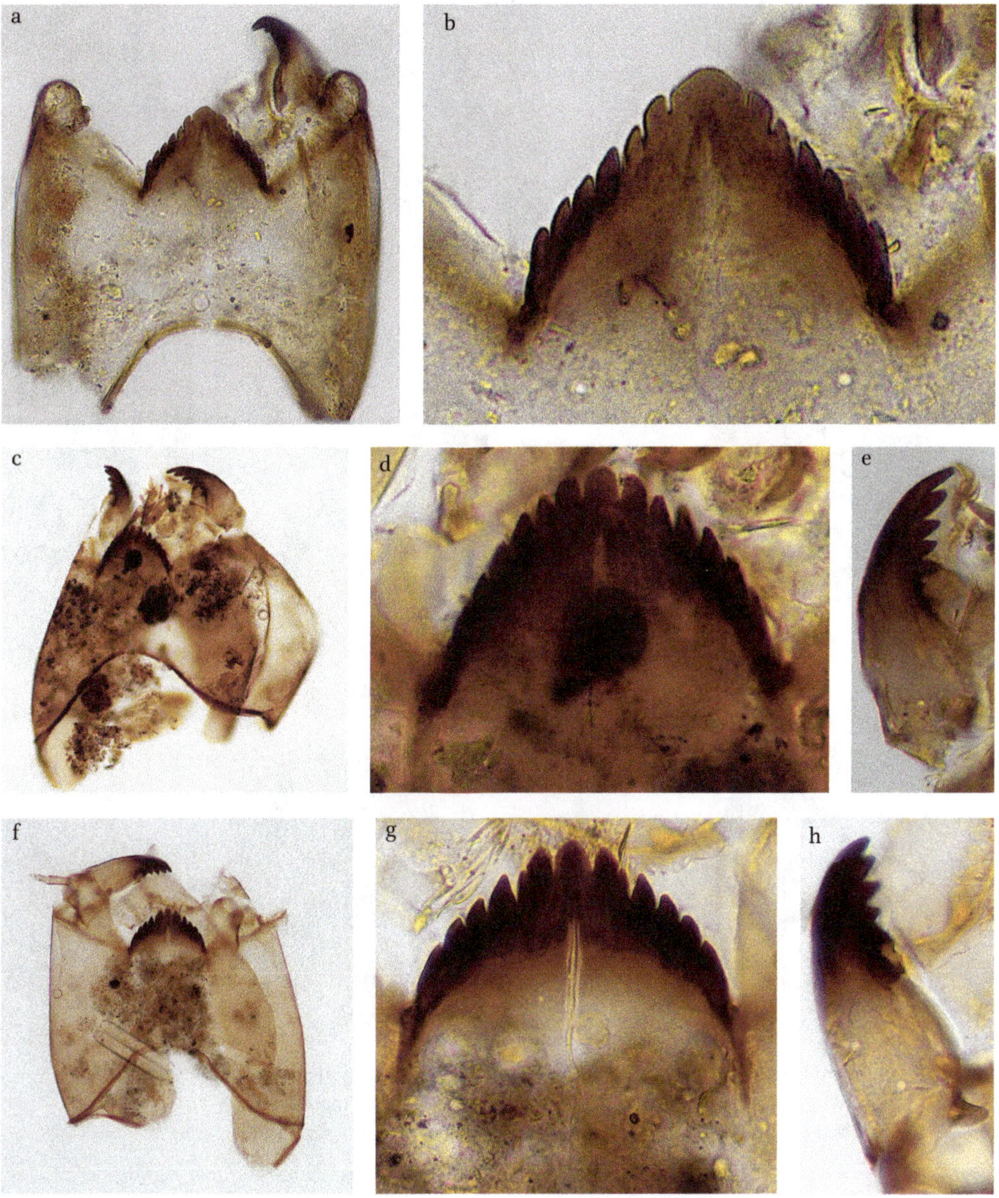

FIGURE 5.5 *Cricotopus. Cricotopus cylindraceus*-type: **a** head capsule, **b** mentum. *Cricotopus tremulus*-type: **c** head capsule, **d** mentum, **e** mandible. *Cricotopus intersectus*-type: **f** head capsule, **g** mentum, **h** mandible

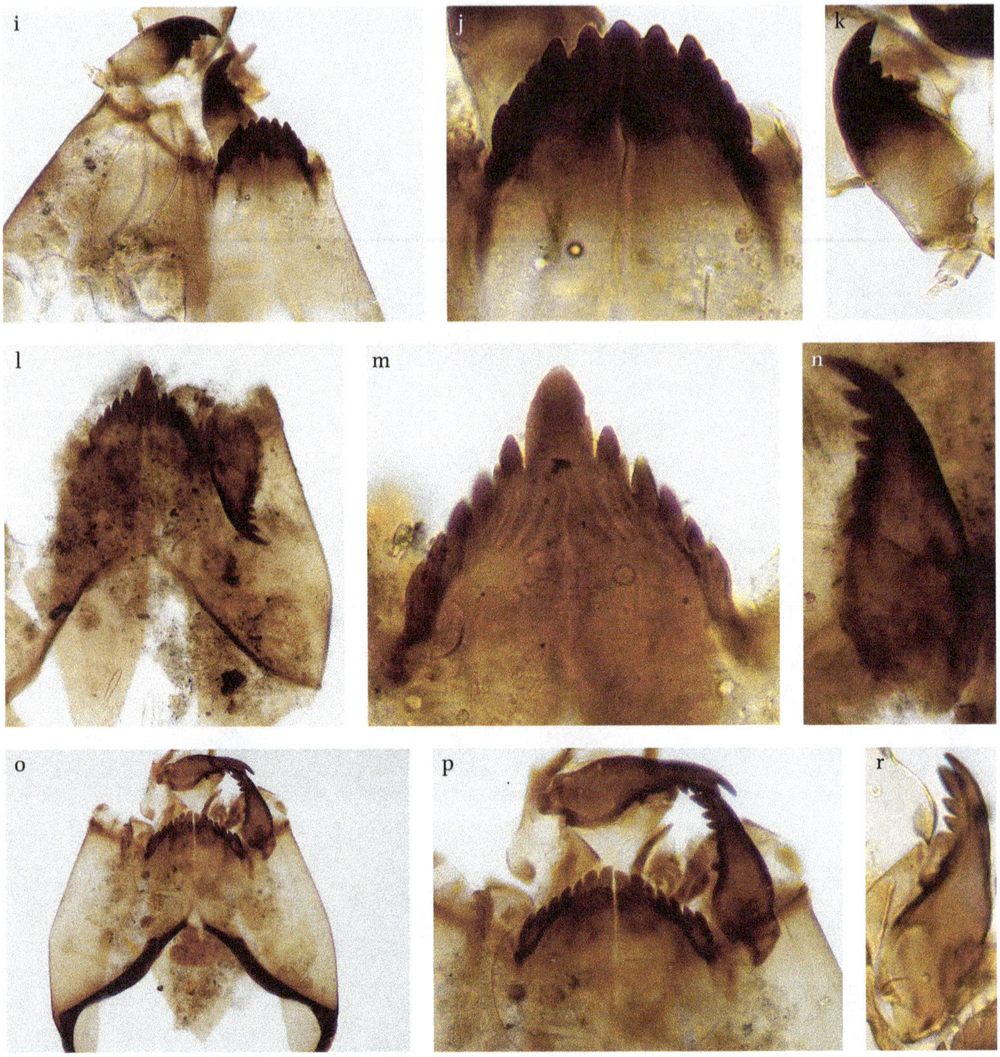

FIGURE 5.5 (continuation) *Cricotopus obnixus*-type: **i** head capsule, **j** mentum, **k** mandible. *Cricotopus sylvestris*-type: **l** head capsule, **m** mentum, **n** mandible. *Cricotopus skirwithensis*-type: **o** head capsule, **p** mentum, **r** mandible

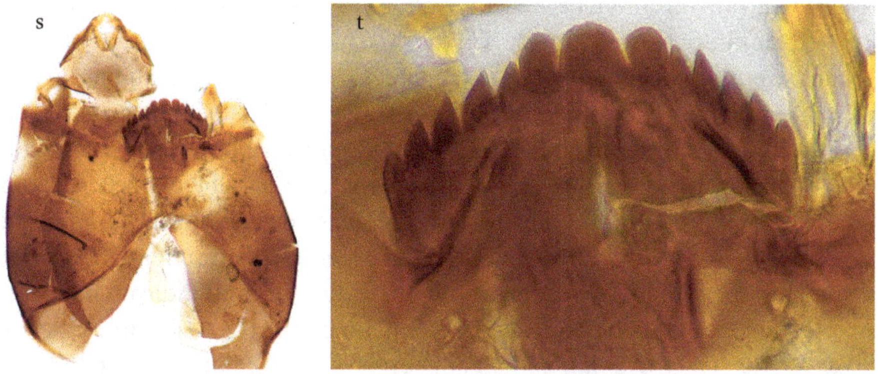

FIGURE 5.5 (continuation) *Cricotopus* type Tatra: s head capsule, t mentum

First lateral tooth of the mentum is clearly rounded, wider in the middle than at the base. Remainder of lateral teeth pointed (Fig. 5.5p). First inner tooth of the mandible large, noticeably broader than apical tooth, but not taller; outer margin of mandible smooth or only with a vague indication of crenulations (Fig. 5.5r). Setae submenti positioned between the bases of ventromental plates. Premandible with one apical tooth (however, there may be a faint indication of a notch).

***Cricotopus* type Tatra.** Head capsule light brown to yellow, occipital margin only slightly darker (Fig. 5.5s). Mentum rather horizontal in shape, all teeth brown. Single median tooth, about 1.5 times width of first lateral tooth; first lateral teeth obviously rounded. Second lateral tooth reduced and partly fused with first lateral tooth. Setae submenti situated below the bases of mentum (Fig. 5.5t). This type corresponds to *Cricotopus/Orthocladius* type I in Chamutiová et al. (2020).

Distribution

Cricotopus remains were relatively frequently recorded in all studied lake districts situated at lower altitudes (except for lakes Morské oko, SK and Vinderel, RO).

Ecology and palaeoecological interpretation

Cricotopus is a species rich genus. Larvae inhabit a wide range of aquatic habitats; however, various species or species groups differ significantly in their ecological requirements. While larvae of subgenera *Cricotopus* and *Paratrichocladius* occur predominantly in cold, well-oxygenated flowing waters, larvae of the *Isocladius* subgenus live mostly in stagnant water bodies, and slowly moving, often organically polluted rivers, and can exist under poor oxygen conditions. As scrapers, they are mostly associated with biofilms on stones, wood and macrophytes, some species mine in living aquatic plants (Hirvenoja 1973, Moller Pillot 2013, Andersen et al. 2013).

Due to issues in larval identification, *Cricotopus* subfossils are grouped to larger taxonomic categories with *Orthocladius*, and possibly with *Paratrichocladius*, in many palaeolimnological studies. Consequently, substantial amount of ecological information may be lost in palaeo-reconstructions. *Cricotopus* remains can be frequent in some sediment sequences and indicate increased lake productivity with developed littoral macrophytes (e.g., Brooks et al. 2001, Brodersen & Quinlan 2006, Langdon et al. 2006, Raposeiro et al. 2017). Cold-stenothermic and rheophilic species of the subgenus *Paratrichocladius* (*C.* (*P.*) *nivalis, C.* (*P.*) *skirwithensis*) occur infrequently in alpine lakes (Boggero et al. 2006, Bitušík et al. 2006) and the same applies for larvae of the *Cricotopus tremulus* group (Novikmec et al. 2015). Larval remains are found sporadically in the sediment records, and can be associated with cold, well-oxygenated littoral of oligotrophic lakes.

Diplocladius Kieffer (Fig. 5.6)

Head capsule dark brown, **occipital margin** black (Fig. 5.6a). **Antenna** 5-segmented, premandible bifid. **Mandible** with 4 inner teeth, their combined width is larger than length of apical tooth. **Mentum** broad with two median teeth and 6 pairs of lateral teeth. First lateral tooth is subequal to median tooth and second lateral tooth is smaller than first and third lateral teeth (Figs. 5.6a,b). **Ventromental plates** are long and broad with noticeably long and dense beard (Fig. 5.6b).

Remarks

The unique shape of mentum with long beard is distinctive for the genus.

Distribution

Remains of *Diplocladius cultriger* Kieffer, 1908, the only species known in the Holarctic were rare in lake sediments, only recorded occasionally from few lakes in the Tatra Mts. (SK) and in one lake in the Chornohora Massif (UA).

Ecology and palaeoecological interpretation

Larval *Diplocadius* are primarily rheophilic, occupying brooks, spring and ditches; records from lakes are rare and come from the littoral of cold lakes

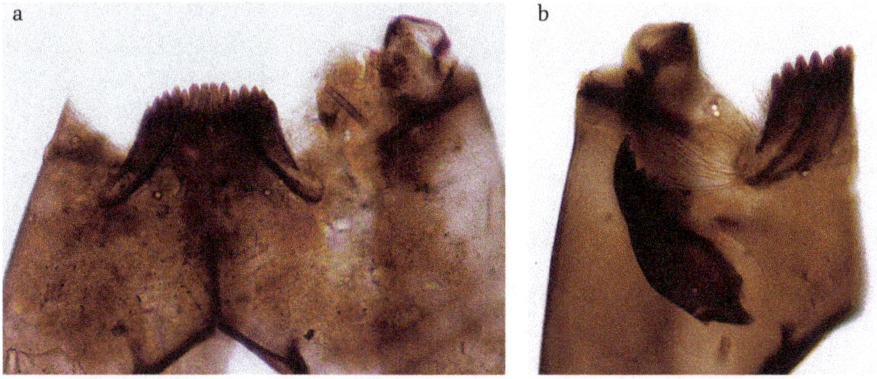

FIGURE 5.6 *Diplocladius*. **a** head capsule, **b** detail of mentum and ventromental plate with long and dense beard

(Moller Pillot 2013, Bitušík et al. 2000). Larvae seem to be tolerant of lower oxygen content; they were collected from various substrates, such as silt, coarse detritus, wood, stones and mosses (Moller Pillot 2013 and references therein).

In the palaeomaterial it indicates presence of a cold inlet stream and/or cold littoral zone.

Epoicocladius Šulc & Zavřel (Fig. 5.7)
Head capsule yellow brownish with brown **occipital margin** (Fig. 5,7a). **Antenna** 4-segmented, blade extending beyond flagellum (Fig. 5.7c). **Premandible** bifid. Apical **mandibular** tooth long, 2–3 inner teeth with only the innermost tooth clearly separated (Fig. 5.7b). **Mentum** strongly arched, apex with median group of 5–7 visible teeth, 4–5 lateral teeth often invisible in uncompresseed heads (Fig. 5.7b). **Ventromental plates** distinct, beard absent (Fig. 5.7b).

Remarks
The shape of mentum makes *Epoicocladius* unique among orthoclads.

Distribution
Subfossil remains were very rare, only recorded from Morské oko lake, SK (due to the occurrence of the mayfly *Ephemera danica* in the littoral zone on which larvae live phoretically, i.e., attached for the purpose of travel). Living larvae of *Epoicocladius* were found up to 5 m depth, however their maximum abundance was at 2 m (unpublished data).

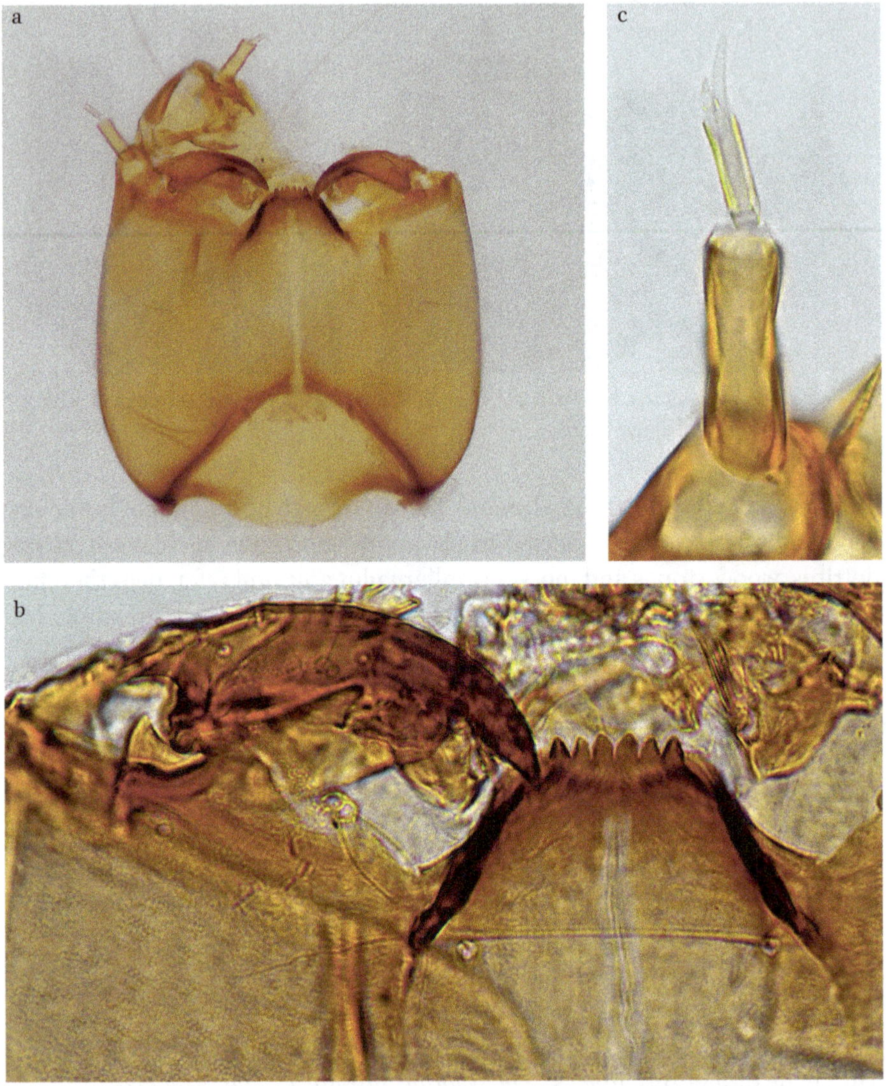

FIGURE 5.7 *Epoicocladius*. **a** head capsule, **b** mentum and mandible, **c** antenna

Ecology and palaeoecological interpretation

Larvae of the single Palaearctic species, *E. ephemerae* (Kieffer, 1924) live phoretically on reophilic mayfly larvae of the genus *Ephemera* (Soldán 1988), consequently their occurrence in stagnant waters is rare. Like its host, larval *Epoicocladius* prefer sandy substrates with coarse and fine detritus in unpolluted or slightly polluted brooks and streams (Moller Pillot 2013). Our data indicate that larvae can also occupy the littoral zone of unpolluted lakes if sediment and oxygen content are sufficient for the host.

Larval remains indicate well oxygenated littoral with sandy substrate and high abundance of *Ephemera* larvae. It can also refer to the presence or increased discharge of an inlet stream.

Eukiefferiella Thienemann (Fig. 5.8)

Head capsule narrow, yellow to dark brown. **Occipital margin** darkened, broad (Figs. 5.8a,d,f,h). **Antenna** with 4–5 segments, **premandible** broad with one apical tooth. **Mandible** with 3–4 inner teeth, apical tooth shorter than combined width of inner teeth; mola with 1–5 spines (Figs. 5.8c,g,j). **Mentum** striped, with 1 or 2 median and 4–5 lateral teeth (Figs. 5.8b,e,g,i,j). **Ventromental plates** narrow, creating an appearance of a dark triangle below the last lateral tooth of mentum (Figs. 5.8b,e,g,i).

Remarks

Eukiefferiella is very similar to *Tvetenia* (Fig. 5.27) and the most reliable character, separating their subfossils is the position of setae submenti, which are close to the base of the mentum in *Eukiefferiella* but well below the mentum in *Tvetenia*. Moreover, head capsule of *Tvetenia* is usually pale, while that of *Eukiefferiella* is light brown to dark brown (Schmid, 1993). Another useful diagnostic feature to distinguish the genus from other genera (but *Tvetenia*) is that mola (the inner part of mandible) bears 1–5 obvious spines (well visible at recent individuals and sometimes on remains, too).

Morphotypes

Four types were distinguished.

1 Mentum with single median tooth 2
– Mentum with double median tooth .... *Eukiefferiella claripennis*-type T

2 Median mental tooth broad, at least 3× as broad as first lateral tooth 3
– Medial mental tooth narrow, at most 2× as broad as firs lateral tooth ....
.... *Eukiefferiella brevicalcar*-type

3 Mentum with 4 lateral teeth .... *Eukiefferiella devonica*-type
– Mentum with 5 lateral teeth .... *Eukiefferiella fittkaui*-type

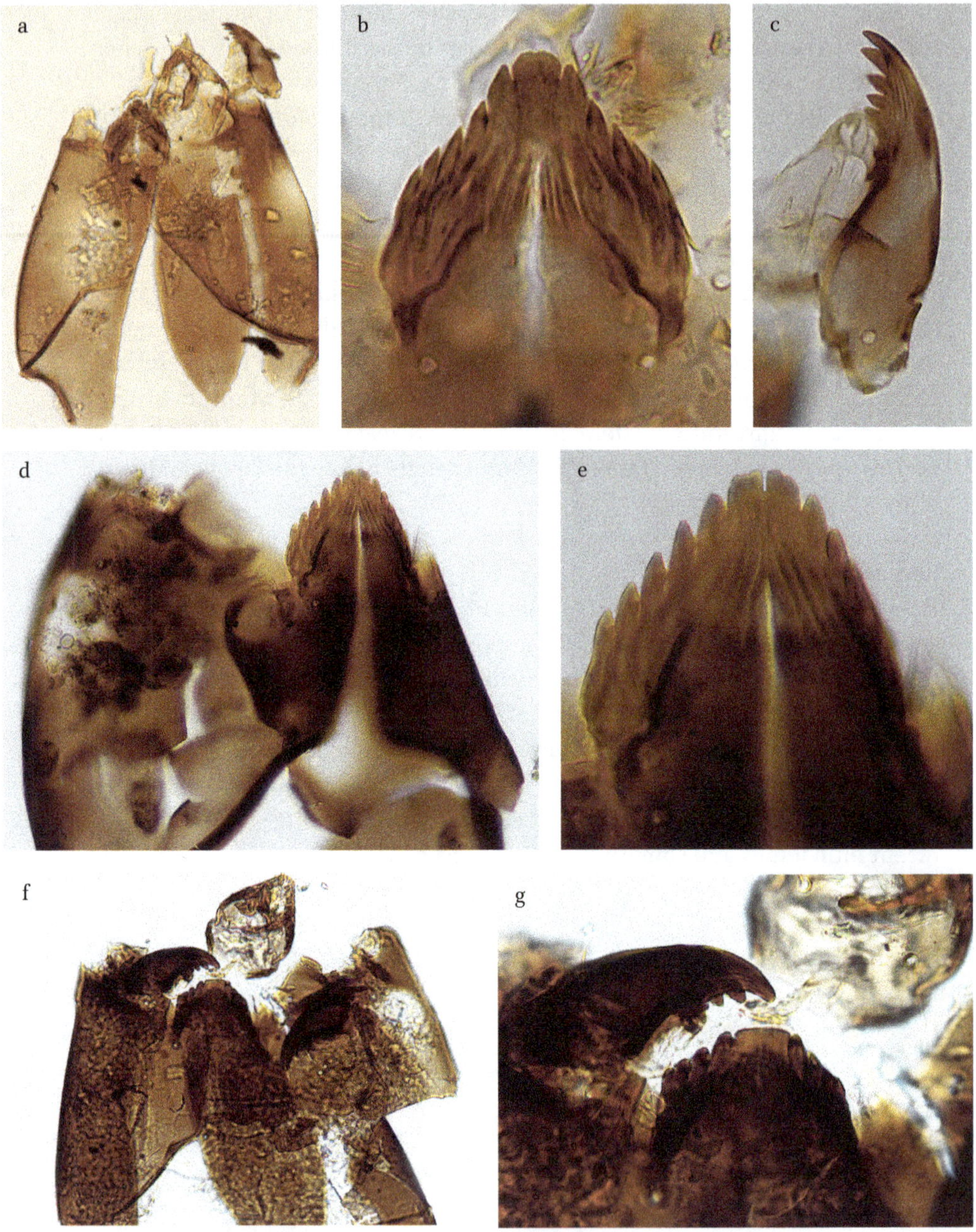

FIGURE 5.8 *Eukiefferiella. Eukiefferiella brevicalcar*-type: **a** head capsule, **b** mentum, **c** mandible. *Eukiefferiella claripennis*-type T: **d** head capsule, **e** mentum. *Eukiefferiella devonica*-type: **f** head capsule, **g** mentum and mandible

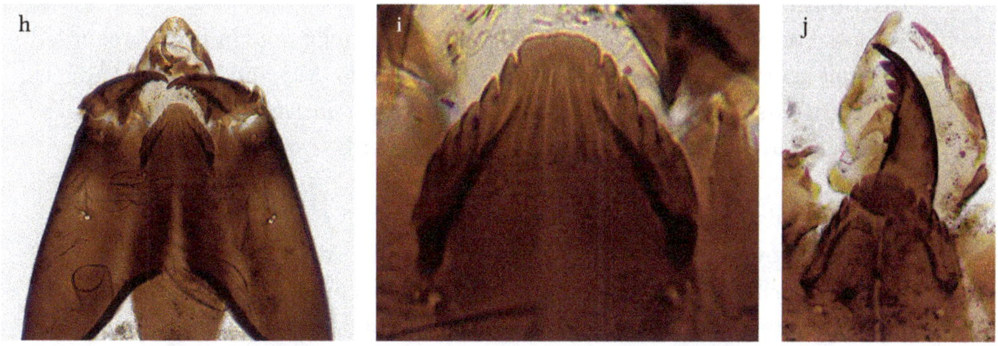

FIGURE 5.8 (continuation) *Eukiefferiella fittkaui*-type: **h** head capsule, **i** mentum, **j** mentum and mandible

***Eukiefferiella brevicalcar*-type.** Head capsule brown (Fig. 5.8a). Mentum with a single median tooth, about twice the width of first lateral tooth. Second lateral tooth pressed to first lateral, leaving an obvious gap between the second tooth and the group of three outermost lateral teeth (Fig. 5.8b).

***Eukiefferiella claripennis*-type T** (T is referring to the *Tokunagaia* genus). Head capsule dark brown (Fig. 5.8d). Mentum with two median teeth and 5 pairs of equidistant lateral teeth (Fig. 5.8e). Brooks et al. (2007) present only one *Eukiefferiella* morphotype with double median tooth, *E. claripennis*-type, which differs in some minor but obvious features from the type found in the Tatra Mts. *Eukiefferiella claripennis*-type sensu Brooks et al. (2007) has narrow and weakly separated median teeth, and first lateral teeth appressed to median teeth. In our case the median teeth were well-separated and broader than first laterals that are clearly separated from the median teeth with a deep notch; this combination of characters reminds of that in *E. rectangularis* group, recently placed to *Tokunagaia rectangularis* group. We assume that our morphotype represents this taxon, occurring in cold springs (Bitušík 2000) and lakes (Hamerlík et al. 2017) in the High Tatra Mts. and sediment cores from the alpine Nižné Terianske pleso lake (Bitušík & Kubovčík 1999). To distinguish the two morphotypes, we named the one recorded in the Tatra lakes *E. claripennis*-type T.

***Eukiefferiella devonica*-type.** Dark brown head capsule (Fig. 5.8f). Strongly pigmented mentum with a single broad median tooth, 2–3 times width of first lateral tooth and 4 pairs of lateral teeth (Fig. 5.8g).

***Eukiefferiella fittkaui*-type.** Head capsule brown (Fig. 5.8h). Mentum with a single, broad median tooth more than 4 times width of first lateral tooth and 5 pairs of lateral teeth (Figs. 5.8i,j).

Distribution

Subfossil remains were relatively frequent but usually not abundant, recorded from all lake districts, except for lakes Morské oko (SK) and Vinderel (RO). In the Tatra Mts. lakes (SK, PL), remains occurred relatively frequently in lakes situated at lower altitudes, especially in those with strong inlet streams.

Ecology and palaeoecological interpretation

It is a specious genus (more than 20 species in Europe, Spies & Sæther 2013) with primarily rheophilic/rheobiontic larvae occupying virtually all types of flowing waters; some species can inhabit littoral of cold, oligotrophic lakes (Janecek & Moog 2002, Bitušík & Hamerlík 2003). Larvae are associated with coarse habitats, such as stones, especially when overgrown with algae, or wood, where they scrape algae or feed on detritus, bacteria and fungi colonies (Moller Pillot 2013). Species are oligostenothermic to eurythermic.

Remains indicate intense inlet and/or well oxygenated, open littoral with coarse sediments in cold lakes (e.g., Hamerlík et al. 2016).

Gymnometriocnemus Edwards/***Bryophaenocladius*** Thienemann (Fig. 5.9)
Head capsule yellow to light brown, **occipital margin** pale to dark (Fig. 5.9a). **Antenna** with 5 segments of variable lengths. **Premandible** with 3 teeth. Apical **mandibular** tooth shorter than combined width of 3 inner teeth. **Mentum** dark brown with paired, relatively broad and rounded median teeth and 4 pairs of lateral teeth gradually decreasing in size (Fig. 5.9b). **Ventromental plates** distinct, sclerotized, sometimes extending beyond outer lateral tooth, resembling ears and visible at the first sight (Fig. 5.9b).

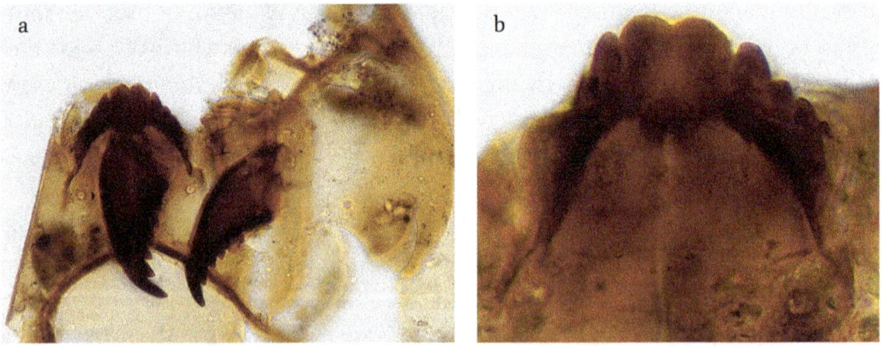

FIGURE 5.9 *Gymnometriocnemus/Bryophaenocladius*. **a** head capsule, **b** mentum and ventromental plates

Remarks

Without body and antennae, larval *Gymnometriocnemus* virtually cannot be distinguished from *Bryophaenocladius*.

Distribution

Subfossils were recorded in low abundance but relatively frequently in the Tatra Mts. lakes (SK, PL) and in one lake in the Chornohora Massif (UA).

Ecology and palaeoecological interpretation

Larvae of *Gymnometriocnemus* and most species of *Bryophaenocladius* are known to be terrestrial, however, a few *Bryophaenocladius* species are semiaquatic or perhaps truly aquatic (Bitušík 2000, Moller Pillot 2013). Aquatic records come from brooks and streams but also from the littoral of mountain lakes and ponds. Larvae are frequently recorded from wet mosses.

Presence of remains in the palaeomaterial may indicate erosion events (Brooks et al. 2007), but also, as in case of other terrestrial/semiterrestrial taxa, lake level oscillations (Hamerlík et al. 2017).

Heterotrissocladius Spärck (Fig. 5.10)

Head capsule yellow, **occipital margin** pale or dark (Fig. 5.10a). **Antenna** 7-segmented, seventh segment vestigial. **Premandible** with two indistinct teeth. Apical **mandibular** tooth shorter than combined width of 3–4 inner teeth (Fig. 5.10d). **Mentum** with two (rarely one) median teeth (sometimes accessory teeth are present) and 5 pairs of lateral teeth. First and second lateral teeth clearly longer than other three lateral teeth (Fig. 5.10b). **Ventromental plates** distinct, bulbous, extended beyond margin of mentum (Fig. 5.10c).

Remarks

Heterotrissocladius is similar to *Paratrissocladius*, however, the latter has only 4 pairs of lateral mental teeth (5 pairs in *Heterotrissocladius*). Subfossil remains without antennae and labro-epipharyngeal region could be confused with *Psectrocladius* (Fig. 5.21) and *Chaetocladius* (Fig. 5.3) that have distinct ventromental plates. However, the outmost lateral tooth is usually reduced in *Chaetocladius* (Fig. 5.3e), and ventromental plates of obviously triangular shape with weak beard are present in *Psectrocladius* (Fig. 5.21h).

Morphotypes

One morphotype was distinguished in the subfossil material following Brooks et al. (2007).

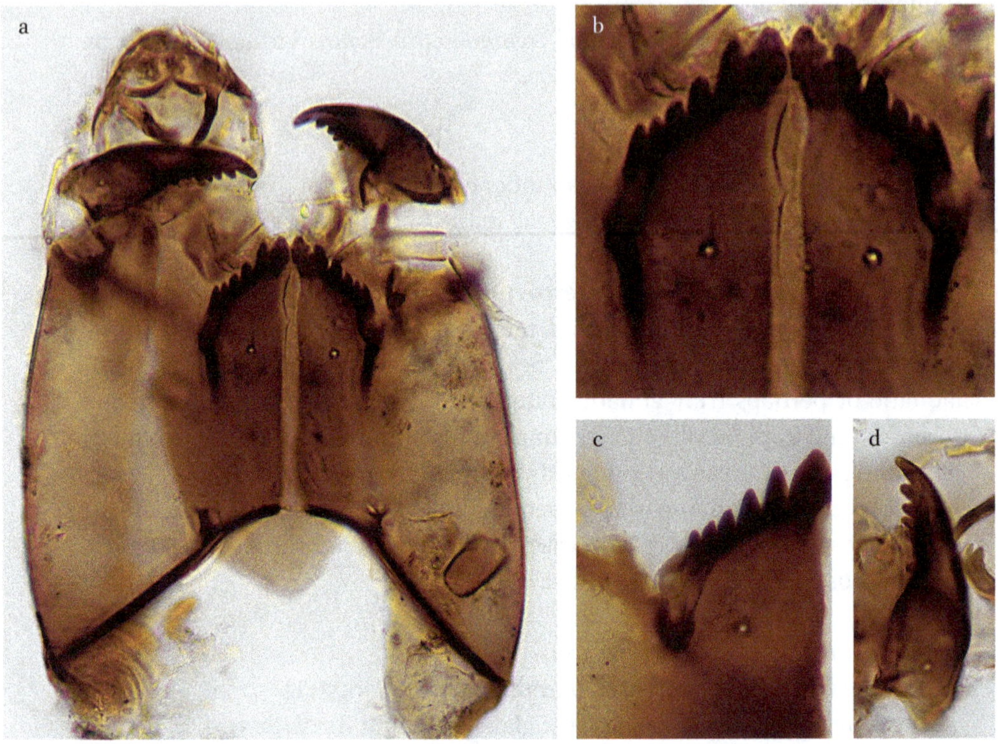

FIGURE 5.10 *Heterotrissocladius marcidus*-type: **a** head capsule, **b** mentum, **c** detail of mentum and ventromental plate, **d** mandible

***Heterotrissocladius marcidus*-type** (Fig. 5.10a). Mentum with two median teeth (Fig. 5.10b). Submentum entirely pigmented, darker than the rest of the head capsule (Fig. 5.10a).

Distribution

Remains were common in the Tatra Mts. lakes (SK, PL); in the Ukrainian Carpathians the morphotype was restricted to the coldest lake in the Chornohora Massif (UA).

Ecology and palaeoecological interpretation

In general, larval *Heterotrissocladius* are oligostenotherms and live both in running and stagnant waters where they can inhabit the whole vertical gradient of the bottom (Bitušík 2000, Janecek & Moog 2002, Moller Pillot 2013). In Slovakia only *H. marcidus* (Walker, 1856) has been recorded, which is the thermally least restricted (Sæther 1975) and most common species distributed throughout Europe (Spies & Sæther 2013). Larvae of *H. marcidus* live in fast-flowing streams, usually in mountain areas, and in the littoral, and, to a lesser extent in profundal of oligotrophic lakes. Larvae are acid tolerant and prefer soft organic sediments (Moller Pillot 2013 and references therein), however, Hamerlík & Bitušík (2009) found significantly higher proportion of *H. marcidus* in mineral sediment (stones) relative to soft organic rich sediments in the littoral of an alpine Tatra Mts. lake situated above 2000 m a.s.l.

The presence of *Heterotrissocladius* remains in the subfossil material will indicate extensive, well oxygenated, cold littoral zone or strong inlet influence.

Hydrobaenus **Fries (Fig. 5.11)**

Head capsule yellow to light brown, **occipital margin** dark brown (Fig. 5.11a). **Antenna** 6-segmented, sixth segment vestigial, blade shorter than flagellum. **Premandible** with two apical teeth. **Mandible** with 3 inner teeth, apical tooth rather long. **Mentum** with two median teeth (in worn down specimens may look like one broad tooth) and 6 pairs of lateral teeth. Median teeth taller than first lateral teeth (Fig. 5.11b). **Ventromental plates** rounded and extending beyond outer lateral mental tooth (Fig. 5.11c).

Remarks

Hydrobaenus is similar to *Oliveridia* (not shown in this book), *Trissocladius* (not shown in this book) and *Zalutschia* (Fig. 5.28). Mentum of *Oliveridia* has one median tooth. Median teeth of mentum in *Trissocladius* are lower than first lateral teeth. In *Zalutschia* median teeth are wider, first lateral teeth are reduced and there is a weak beard at the base of mentum (beard is absent in *Hydrobaenus*).

Morphotypes

One morphotype was distinguished in the subfossil material following Brooks et al. (2007).

***Hydrobaenus conformis*-type** (Fig. 5.11a). An important diagnostic character of this morphotype is the inwardly angled outer margin of median teeth at about mid-height. Moreover, lateral teeth are subequal, as opposed to the very narrow first two lateral teeth in *H. lugubris*-type (not shown) (Fig. 5.11b), and ventromental plates apically more truncate than bulbous (Fig. 5.11c).

Distribution

The morphotype was found exclusively in a dystrophic forest lake in the Tatra Mts. (SK).

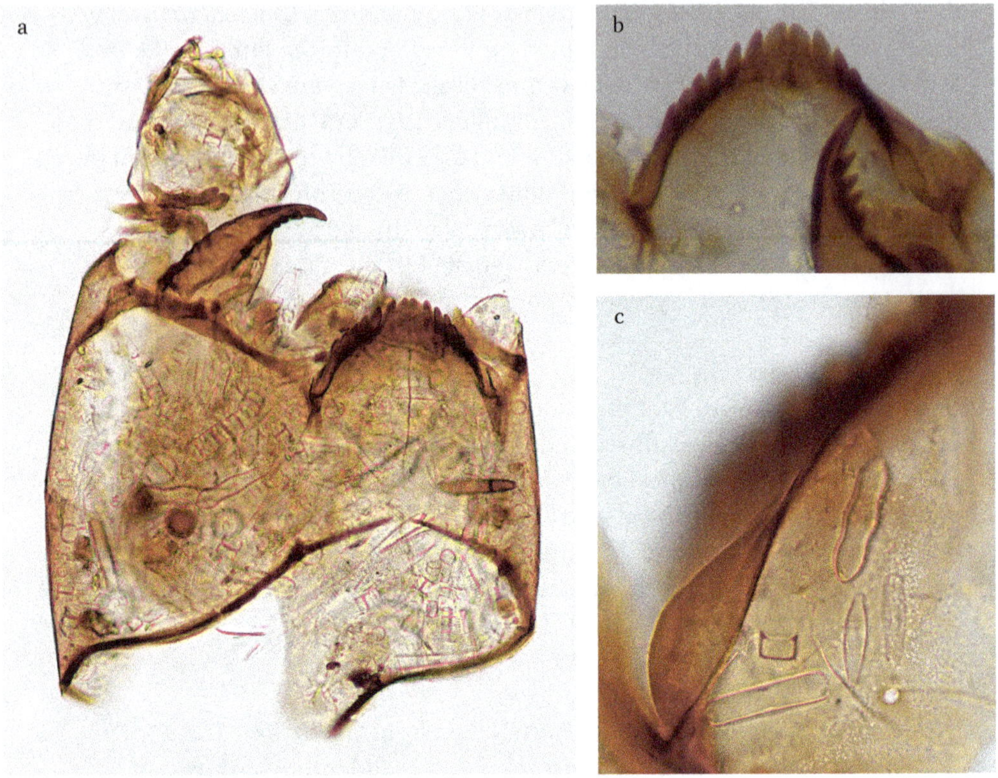

FIGUERE 5.11 *Hydrobaenus conformis*-type: **a** head capsule, **b** mentum and mandible **c** detail of mentum showing rounded ventromental plate

Ecology and palaeoecological interpretation

Hydrobaenus larvae live in the littoral zone of lakes, in small stagnant waters and in streams and rivers. They are oligostenothermic and prefer well-oxygenated conditions (Sæther 1976).

In palaeolimnological studies, the genus is classified as cold-temperature indicator (Laroque & Hall 2003) and its presence in stratigraphic records may be connected with oligotrophic conditions (Brooks & Birks 2004).

Limnophyes Eaton/***Paralimnophyes*** Brundin (Fig. 5.12)

Head capsule yellow, **occipital margin** pale or darkened (Fig. 5.12a). **Antenna** 5-segmented. **Premandible** with two apical and two inner teeth. Apical **mandibular** teeth shorter than combined width of 3 (*Limnophyes*) or 4 (*Paralimnophyes*) inner teeth (Fig. 5.12c). **Mentum** with two median teeth slightly higher than first lateral of 5 lateral teeth. Fourth and fifth laterals obviously smaller and narrower than others. Mentum frequently with pale stripes in the median region (Figs. 5.12b,c). **Ventromental plates** widened and sclerotized posterolaterally into a rounded dark pigmented bulge projecting below outermost lateral tooth and appearing as additional basal tooth separated by notch (Figs. 5.12b,c).

Remarks

When mandibles are missing from subfossils, *Paralimnophyes* and *Limnophyes* are indistinguishable (if present, 3 inner teeth will distinguish *Limnophyes* from *Paralimnophyes* with 4 inner teeth). Shape of mentum with two median teeth and 5 pairs of laterals, presence of stripes resembles *Eukiefferiella* (Fig. 5.8), however widely separated median teeth and presence of dark brown bulge at the base of ventromental plates is distinctive for *Limnophyes*/*Paralimnophyes* (Figs. 5.12b,c). *Heleniella* (not shown in this book) has a similar pattern of mentum but unlike in *Limnophyes*/*Paralimnophyes*, the fifth lateral tooth is longer than the fourth and there is a U-shaped gap between median teeth.

Distribution

Remains were found in all lake district except for Morské oko lake (SK). Head capsules were frequent (but not abundant) in most of the lakes in the Chonohora and Svydovets Massifs (UA) and reservoirs in Štiavnické vrchy Mts. (SK); in the Tatra Mts. they were especially common in the PL lakes.

Ecology and palaeoecological interpretation

Larvae of most *Limnophyes* species are terrestric/semiterrestric and only few species are truly aquatic. Larvae can be found in the upper layers of wet bottoms,

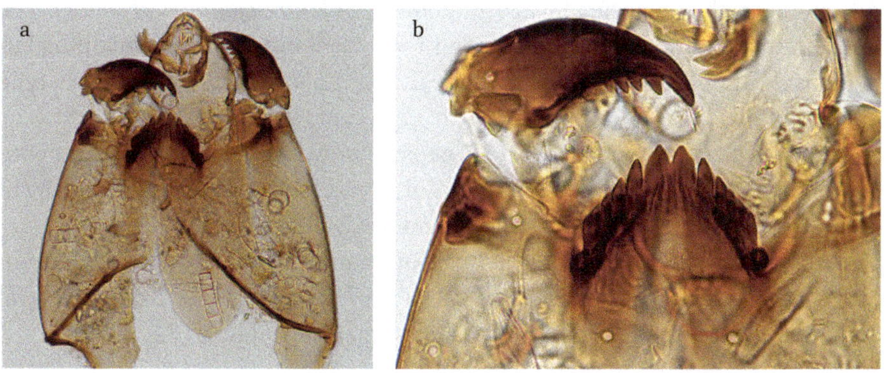

FIGURE 5.12 *Limnophyes/Paralimnophyes*. **a** head capsule, **b** mentum, **c** mentum and mandible. Note that the individual on the picture is *Limnophyes*, since it has a mandible with 3 inner teeth

hygropetric environments, as well as on plants and wood near the water surface of usually smaller stagnant and flowing waterbodies (Moller Pillot 2013). In the Tatra Mts. the genus was typical for smaller lakes/ponds of high altitudes (Hamerlík et al. 2017). Larvae of the closely related *Paralimnophyes* live in small lowland waterbodies and bogs at the water edge and among vegetation (Moller Pillot 2013).

In subfossil samples it can be an indicator of lake level fluctuations or erosion events (Brooks et al. 2007 and references therein).

Metriocnemus van der Wulp (Fig. 5.13)
Head capsule yellow to dark brown, **occipital margin** pale or dark (Figs. 5.13a,d). **Antenna** with 5 segments and variable in structure, sometimes reduced. **Premandible** with 2–4 apical teeth. Apical **mandibular** tooth short, 4–5 inner teeth are present (Figs. 5.13c,f). **Mentum** with 1–4 median teeth, usually shorter than first of 5–6 pairs of lateral teeth (Figs. 5.13b,e). **Ventromental plates** very weak (Figs. 5.13b,e).

Remarks
The characteristic shape of mentum with shorter and narrower median teeth than first laterals is distinctive for the genus, however, the same pattern of mentum is typical for *Thienemannia* (not shown in this book), so heads of both genera are indistinguishable in subfossil material.

Morphotypes
Two morphotypes were distinguished following Brooks et al. (2007).

1 Lateral teeth decreasing in size gradually .... *Metriocnemus eurynotus*-type
– 1st and 2nd lateral teeth much larger than median teeth and remaining laterals .... *Metriocnemus fuscipes*-type

***Metriocnemus eurynotus*-type**. Head capsule brown (Fig. 5.13a). Mentum with two short median teeth and with 5 pairs of lateral teeth gradually decreasing in size, 1st lateral tooth broader than median teeth (Fig. 5.13b).
***Metriocnemus fuscipes*-type**. Head capsule dark brown (Fig. 5.13d). Mentum with two short median teeth and with 5 pairs of lateral teeth, 1st and 2nd lateral teeth taller and broader than median and remaining laterals (Fig. 5.13e).

Distribution
Metriocnemus remains were recorded rarely and in low abundances in all lake districts except for Morské oko lake (SK).

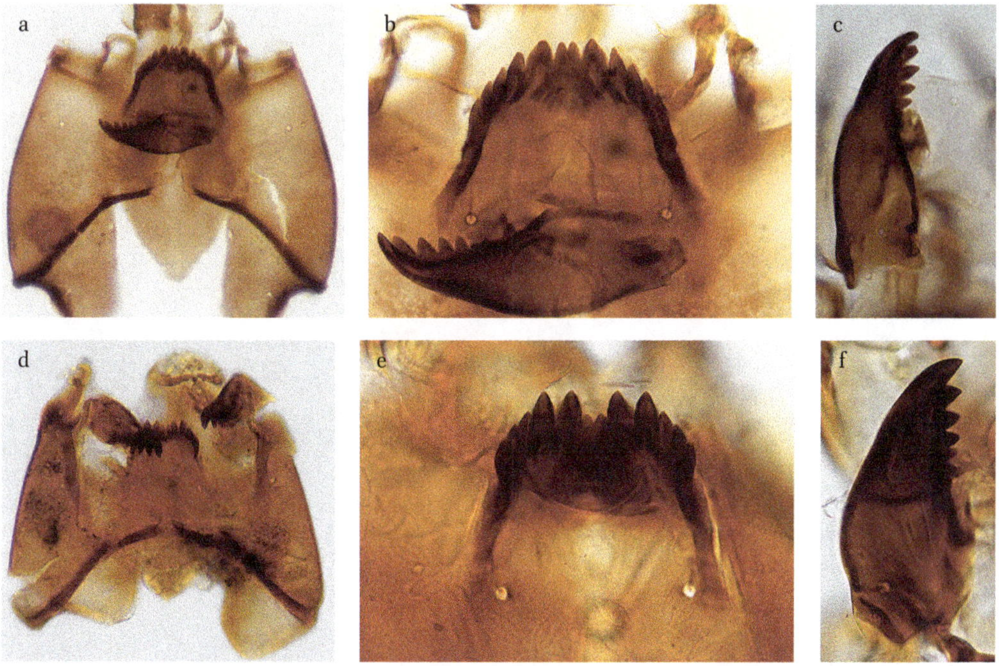

FIGURE 5.13 *Metriocnemus. Metriocnemus eurynotus*-type: **a** head capsule, **b** mentum and mandible, **c** mandible. *Metriocnemus fuscipes*-type: **d** head capsule, **e** mentum, **f** mandible

Ecology and palaeoecological interpretation

Most larval *Metriocnemus* live terrestrially/semiterrestrially in wet soils, mosses, hygropetric habitats and water-filled tree holes (dendrotelmata). Aquatic species are known from small springs and streams, some from the littoral zone of lakes (Janecek & Moog 2002).

Similar to other semiterrestrial taxa, subfossil remains may indicate lake level fluctuations or erosion events in the catchment.

Nanocladius Kieffer (Fig. 5.14)

Head capsule small, pale yellow to yellowish brown, **occipital margin** darker than head capsule (Fig. 5.14a). **Antenna** 5-segmented. **Premandible** with simple or indistinct divided apical tooth, or with 3–5 apical teeth. Apical **mandibular** tooth very long, slender, 3 small inner teeth (Fig. 5.14c). **Mentum** with broad, partially paired median tooth and normally with six pairs of lateral teeth, sometimes only 3 or 5 pairs recognizable or hardly visible (Fig. 5.14b). **Ventromental plates** narrow and long to extremely long, with rounded or straight apex, beard absent in European species (Fig. 5.14b).

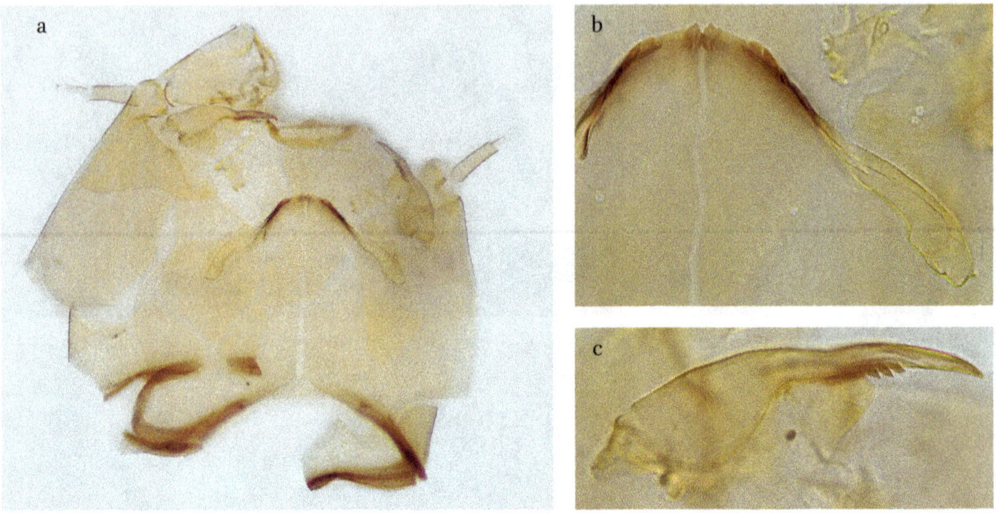

FIGURE 5.14 *Nanocladius rectinervis*-type: **a** head capsule, **b** mentum and ventromental plate, **c** mandible

Remarks

The shape of mentum with remarkably long ventromental plates is distinctive for the genus.

Morphotypes

One morphotype was distinguished.

***Nanocladius rectinervis*-type** (Fig. 5.14a). Ventromental plates long and lateral teeth of mentum well recognizable (Fig. 5.14b), premandible single or slightly bifid. The *N. balticus* morphotype (not shown in this book), which has been found frequently as pupal exuviae in the reservoirs of the Štiavnické vrchy Mts. (Bitušík et al. 2007) is distinctive by extremely long ventromental plates, mentum with poorly defined lateral teeth and premandible with 3–5 teeth.

Distribution

Subfossil remains were rare and restricted to reservoirs in the Štiavnické vrchy Mts. (SK).

Ecology and palaeoecological interpretation

Generally, larvae of *Nanocladius* are rheophilic, however *N. balticus* (Palmén, 1959) lives in lake littoral, and *N. dichromus* (Kieffer, 1906) prefers stagnant waters and slowly flowing river courses. Data from European freshwaters indicate that larvae of both species predominantly occur in mesotrophic and eutrophic conditions, however, according to Moller Pillot (2013) *Nanocladius* species require high oxygen content.

Remains in the palaeomaterial will indicate higher organic content (meso- to eutrophy) but relatively well oxygenated littoral zone.

Orthocladius van der Wulp (Fig. 5.15)
Head capsule yellow, brown to dark brown, or reddish brown, **occipital margin** usually dark (Figs. 5.15d,f). **Antenna** 5-segmented (rarely with 4 segments). **Premandible** with one (frequently notched) or two apical teeth. **Mandible** with 3 inner teeth, apical tooth usually short, longer in some species of subgenera *Euorthocladius* and *Symposiocladius* (Figs. 5.15c,h). **Mentum** with one median and usually with 6 pairs of lateral teeth; sometimes 7–9 laterals are present (Figs. 5.15b,e,g) and *O.* (*Symposiocladius*) *lignicola* has an unusual mentum with single elongate median tooth and 2 pairs of lateral teeth. **Ventromental plates** narrow and short, beard absent (Figs. 5.15b,e,g) (with the exception of subgenus *Pogonocladius*).

Remarks
Subfossil head capsules of *Orthocladius* with 6 pairs of lateral mental teeth and without beard are essentially indistinguishable from *Cricotopus* (incl. *Paratrichocladius*) (Fig. 5.5) as mentioned above (see remarks to *Cricotopus*).

Morphotypes
Three morphotypes were distinguished following Soponis (1977) and Schmid (1993).

1 Mentum with a exceptionally large triangular median tooth and two very short lateral teeth .... *Orthocladius lignicola*-type
– Mentum with median tooth at most slightly larger than first laterals and six lateral teeth 2

2 Head pale, mentum with an eminent median tooth, lateral teeth triangular in shape .... *Orthocladius dentifer*-type
– Head brown, mentum median tooth about as tall as 1st laterals, 2–6 lateral teeth rectangular in shape 3

3 Mentum strongly sclerotized, black, ventromental plates short .... *Orthocladius frigidus*-type
– Mentum weakly sclerotized, brown, ventomental plates long, extended well below seta submenti .... *Orthocladius* (*Euorthocladius*) sp.

***Orthocladius* (*Orthocladius*) *dentifer*-type.** Head capsule yellow to light brown with brown occipital margin and mandibular teeth (Fig. 5.15a). Mentum triangular with clearly separated teeth. Median tooth with first two lateral teeth lighter than the reminding laterals. First lateral teeth clearly pointing outwards (Fig. 5.15b). Mandible with crenulated outer margin, inner teeth subequal in size (Fig. 5.15c).

***Orthocladius* (*Mesorthocladius*) *frigidus*-type.** Head capsule brown with black occipital margin (Fig. 5.15d). Mentum with single median tooth and 6 pairs of lateral teeth, all black. Median and first lateral tooth rounded; first lateral tooth is slightly taller than the median tooth (Fig. 5.15e). Seta submenti is located between the narrow ventromental plates, very close to them (Fig. 5.15e); 2 setal pores present above each other, a bigger anterior and smaller posterior. Outer margin of mandible crenulated.

***Orthocladius* (*Symposiocladius*) *lignicola*-type.** Head capsule yellow to light brown with dark occipital margin (Fig. 5.15i). Mentum with a large broadly triangular median tooth and two small lateral teeth, frequently worn (Fig. 5.15j). Mandible broad, bearing three short inner teeth and an apical tooth. Premandible simple, brush absent. In Brooks et al. (2007) the morphotypes is presented as *Symposiocladius lignicola*; recently it has been merged with the *Orthocladius* genus as its subgenus (Sæther 2004).

***Orthocladius* (*Euorthocladius*) sp.** Head capsule dark brown with black occipital margin (Fig. 5.15f). Mentum triangular with weak striation and light brown teeth. Median tooth and first lateral teeth rounded, first lateral narrower at the base (Fig. 5.15g). Ventromental plates characteristic: darkly pigmented, extended well below seta submenti, angling inwards apically, with an obvious bulb at the base. Setae submenti positioned below the base of ventromental plates (Fig. 5.15g). Mandible with first inner tooth large, outer margin smooth (Fig. 5.15h).

Distribution

Orthocladius was recorded in all lake districts except for lakes Morské oko (SK) and Vinderel (RO) but remains were generally rare and never abundant.

Ecology and palaeoecological interpretation

One of the largest genera of the subfamily with at least 6 accepted subgenera (Sæther 2005) and over 100 Holarctic species (Ashe & O'Connor 2012). Larval *Orthocladius* inhabit virtually all types of inland waters, including extreme habitats, such as thermal waters, submerged wood (in which they are mining), hygropetric zone, moist soils, and temporary ponds (Andersen et al. 2013). Some subgenera (e.g., *Eudactylocladius, Euorthocladius* and *Mesorthocladius*)

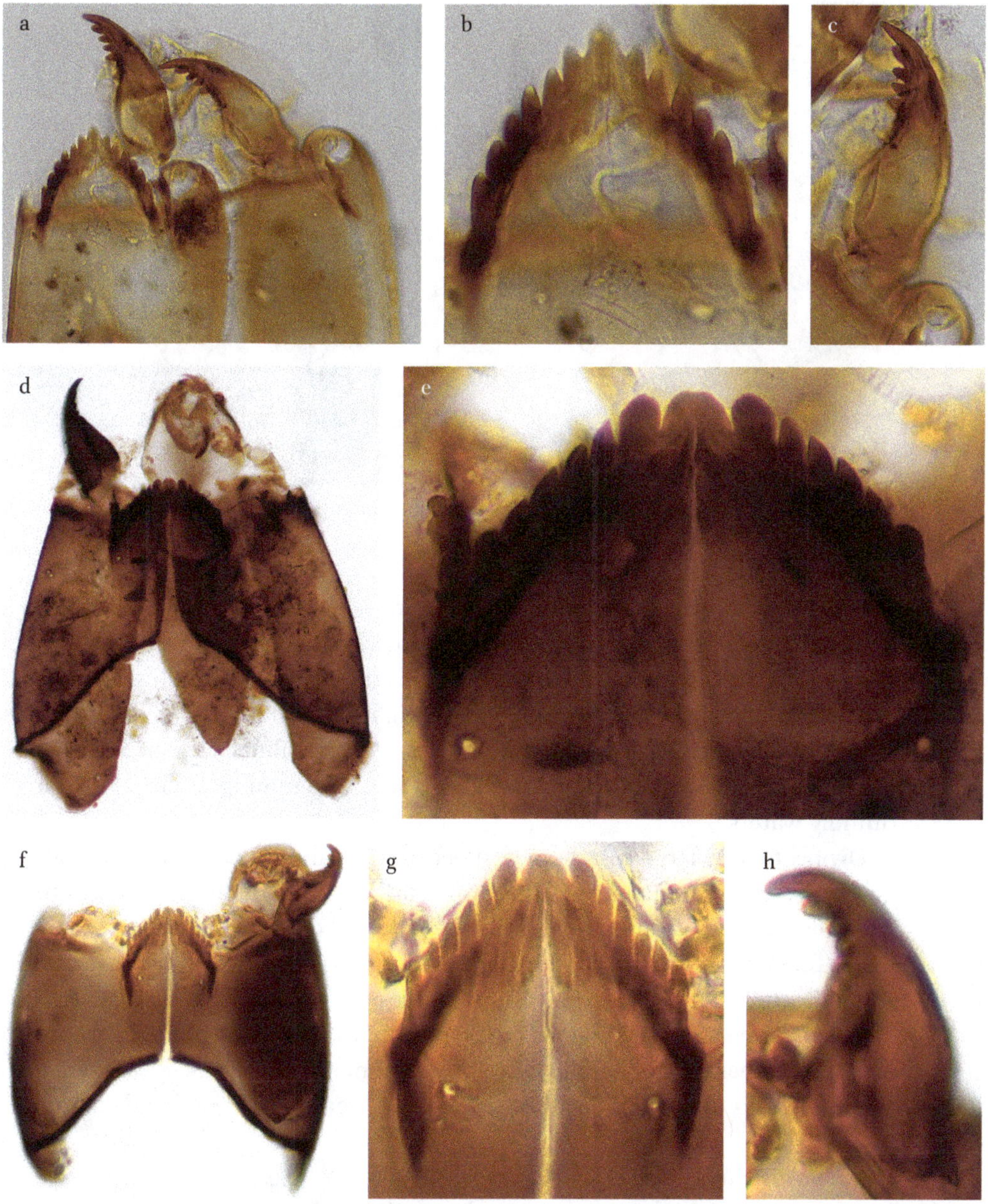

FIGURE 5.15 *Orthocladius*. *Orthocladius dentifer*-type: **a** part of head capsule, **b** mentum, **c** mandible. *Orthocladius frigidus*-type: **d** head capsule, **e** mentum. *Orthocladius* (*Euorthocladius*) sp.: **f** head capsule, **g** mentum, **h** mandible

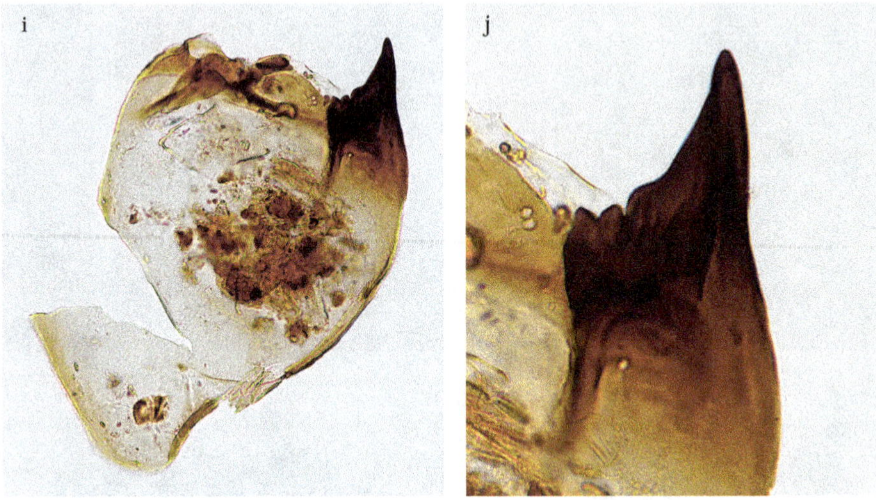

FIGURE 5.15 (continuation) *Orthocladius lignicola*-type: **i** head capsule (half), **j** mentum

are cold-stenothermal. Larvae of *Orthocladius frigidus* primarily inhabit fast-flowing stretches of streams (Moller Pillot 2013), but have been recorded from springs, lowland streams, and littoral of cold lakes (below 10 °C, Rossaro 1991) as well. In Slovakia, the species reaches highest abundances in mountainous streams but can also be found in lower sections (Bitušík 2000, Bitušík & Hamerlík 2003). Larvae of *Orthocladius lignicola* are mining in dead wood in running waters.

Owing to the large number of species and variable ecological requirements, subfossil remains may indicate various environmental conditions. Palaeoecological interpretation should be made carefully and according to the level of identification of the remains. All morphotypes recorded in our study will indicate oligotrophic to mesotrophic lake conditions and some types (e.g., *O. frigidus*-type and *O.* (*Euorthocladius*)) will point to fast-flowing oligosaprobic inflow streams (Moller Pillot 2013). Larval remains of *O. lignicola* are rare in lake sediments and will indicate forested inlet stream with rich woody debris.

Paracladius Hirvenoja (Fig. 5.16)
Head capsule light brown, **occipital margin** darker (Fig. 5.16a). **Antenna** 5-segmented, premandible bifid. **Mandible** with 3 inner teeth, apical tooth slender and long (Fig. 5.16b). **Mentum** with a pale broad and short median tooth, and six pairs of dark lateral teeth (Fig. 5.16a). **Ventromental plates** narrow, projecting beyond outer lateral teeth of mentum, beard present but often indistinct.

Remarks

Shape of mentum with broad and pale median tooth combined with elongated ventromental plates and sparse beard (not always visible) is distinctive.

Morphotypes

The morphotype recorded, *Paracladius* sp., was not assigned to any species/type. However, based on the locality where the specimen was found it belongs to *P. conversus* (see comment below).

Distribution

Subfossil remains were rare, only found in one reservoir of Štiavnické vrchy Mts. (SK).

Ecology and palaeoecological interpretation

Larvae of *Paracladius* are known both from flowing and standing waters. The most common European species of the genus, *P. conversus* (Walker 1856) and *P. alpicola* (Zetterstedt, 1850), have considerably different ecological requirements. Larvae of *P. conversus* are eurythermal and tolerant to low oxygen concentration (Bitušík 2000), often recorded in heavily polluted waters, however, they do not tolerate low pH. Larvae are known from slow flowing streams, brooks and ditches, as well as from rhitral and potamal stretches of rivers where they occupy soft sediments (Bitušík 2000, Moller Pillot 2013 and references therein). Larvae are also recorded from muddy and silty bottoms of lake littorals (Bitušík 2000). According to Moller Pillot (2013), the species is known

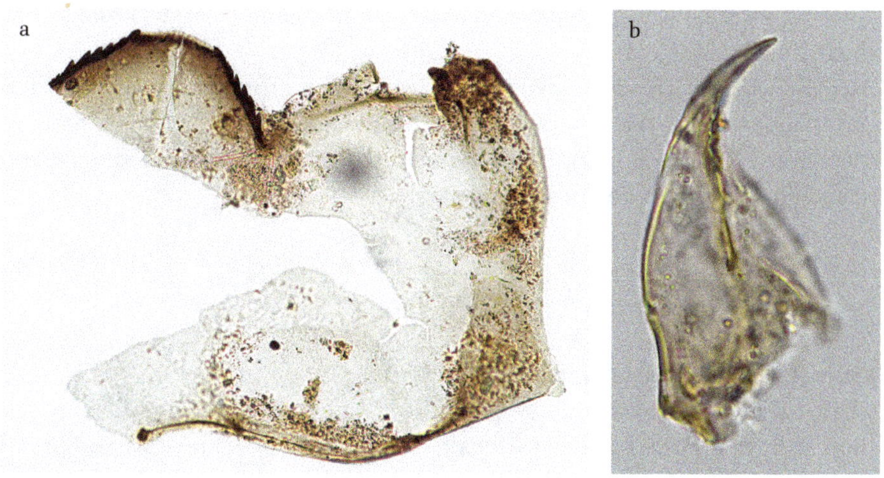

FIGURE 5.16 *Paracladius*. **a** fragment of head capsule, **b** mandible

only from water bodies fed by groundwater. *P. alpicola* (Zetterstedt, 1850) is characteristic for cold northern and alpine lakes (Langton 1991).

Owing to their different ecology, for palaeoecological interpretation it is important to identify which species is recorded. While *Paracladius conversus* will indicate higher trophy and warmer conditions, *P. alpicola* is an indicator of cold, oligotrophic conditions, as it is in the Scandinavian training set, where it is abundant in sediment sequences and in the coldest lakes (identified as *Paracladius* sp., Brooks et al. 2007). The two species/forms, are, however, undistinguishable in larval stage.

Parakiefferiella Thienemann (Fig. 5.17)
Small brownish **head** capsule with pale **occipital margin** (Fig. 5.17a). **Antenna** 6-segmented, sixth segment hair-like, hardly visible. **Premandible** with single or bifid apical tooth. Apical **mandibular** tooth subequal or longer than combined width of 3 inner teeth (Fig. 5.17b). **Mentum** with single median tooth and six pairs of lateral teeth. First laterals often closely adpressed to median tooth, or apex of mentum may be worn, consequently only five pairs of lateral teeth are visible (Figs. 5.17b,c). **Ventromental plates** distinct, usually not extending beyond outermost lateral teeth (Fig. 5.17b).

Remarks
Combination of the single median tooth, six pairs of lateral teeth and well developed ventromental plates is distinctive for *Parakiefferiella*.

Morphotypes
One morphotype was distinguished following Hofmann (1971) in Brooks et al. (2007).
***Parakiefferiella bathophila*-type** (Fig. 5.17a). Apex of mentum pale, first lateral tooth reduced, the other lateral teeth subequal and darkly pigmented. In specimens with worn mentum the median tooth may be rounded and the first lateral tooth may not be visible (Figs. 5.17b,c).

Distribution
Remains were found in all reservoirs in the Štiavnické vrchy Mts. (SK).

Ecology and palaeoecological interpretation
Larvae of *Parakiefferiella* can be found in a wide variety of aquatic biotopes, from springs to streams and various types of standing waters. Larvae prefer sandy bottoms but are also known to live on stones and organic sediments with and without vegetation (Moller Pillot 2013). *P. bathophila* is known from

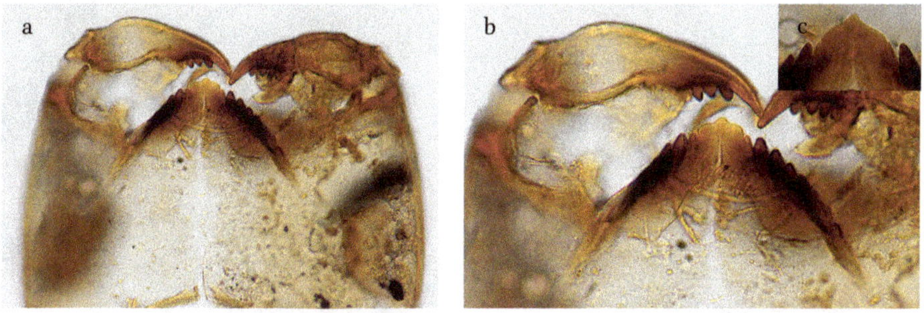

FIGURE 5.17 *Parakiefferiella bathophila*-type: **a** head capsule, **b** mentum and mandible, **c** alternative shape of top of mentum

rhitral and potamal river sections and littoral of lakes; profundal records are known as well (Janecek & Moog 2002).

P. bathophila-type will indicate relatively warm and productive lake conditions (Brooks et al. 2007, Eggermont & Heiri 2011).

Parametriocnemus Goetghebuer/***Paraphaenocladius*** Thienemann (Fig. 5.18)
Head capsule yellow, **occipital margin** slightly darker (Fig. 5.18a). **Antenna** usually 5-segmeted, when 6-segmented, last segment vestigial. **Premandible** with 2–6 teeth. **Mandible** with 3 inner teeth, apical tooth short (Fig. 5.18d). **Mentum** with paired median teeth and five pairs of lateral teeth, fifth lateral minute (Fig. 5.18b). **Ventromental plates** distinct, extending beyond the last lateral tooth, with second plate inside main plate (Fig. 5.18c). **Setae submenti** are positioned high on mentum, about at the level of the outermost lateral teeth (Fig. 5.18b).

Remarks

Subfossil larval head of *Parametriocnemus* essentially cannot be distinguished from *Paraphaenocladius*. If premandibles are present, specimens with two and more than three apical teeth can be assigned to *Parametriocnemus* as premandible of *Paraphaenocladius* bears 3 apical teeth. Our experience is, though, that the shape of ventromental plates, which is clearly foot-shaped in *Parametriocnemus* and more bulbous in *Paraphaenocladius*, may be reliable to distinguish the two genera. Also, most *Paraphaenocladius* species are (semi)terrestrial, and so their occurrence in lake sediments is less likely.

Limnophyes (Fig. 5.12) superficially resembles *Parametriocnemus/Paraphaenocladius* but it differs from it by a darkly pigmented lobe below outermost lateral tooth, subequal lateral teeth, and weakly developed ventromental plates.

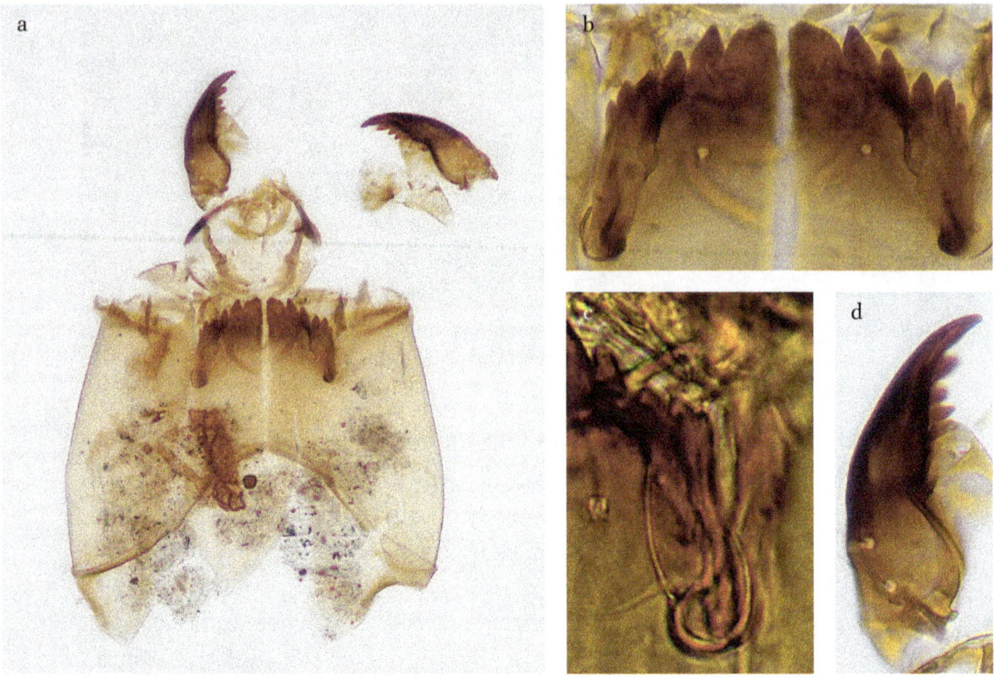

FIGURE 5.18 *Parametriocnemus/Paraphaenocladius*. **a** head capsule, **b** mentum (note the position of seta submenti), **c** detail of mentum showing ventromenatal plate, **d** mandible

Morphotypes

One morphotype, *Parametriocnemus/Paraphaenocladius* was recorded.

Distribution

Subfossil remains were found rarely and in small numbers in almost all studied lake districts expect lakes in the Svydovets Massif (UA) and the Maramures Mts. (RO).

Ecology and palaeoecological interpretation

Larvae of *Parametriocnemus* are rheophilic, inhabiting springs and courses of streams with slow-flowing water. Most *Paraphaenocladius* species are terrestrial and semi-terrestrial, with few truly aquatic species inhabiting springs and streams (Andersen et al. 2013b).

Subfossil remains may indicate lake level fluctuations or erosion events (*Paraphaenocladius*) or more intense inlet stream (*Parametriocnemus*). While according to Brooks et al. (2007) subfossils are frequent in lake sediments, they were rarely encountered in the Carpathian lakes.

Paratrissocladius Zavřel (Fig. 5.19)

Head capsule light brown with brown submentum, **occipital margin** black (Fig. 5.19). **Antenna** with 7 segments, seventh segment hair-like, vestigial. **Premandible** with single apical tooth (may be apically slightly notched) and a low inner tooth. **Mandible** with relatively short apical tooth and 3 inner teeth. **Mentum** with double median teeth and 4 pairs of lateral teeth, fourth as tall as third (Fig. 5.19). **Ventromental plates** moderately developed, extending beyond outermost laterals.

Remarks

Head capsule resembles *Heterotrissocladius* (Fig. 5.10) and can be distinguished by the number of lateral mental teeth: 4 pairs in *Paratrissocladius* as opposed to 5 pairs in *Heterotrissocladius*.

Morphotypes

One morphotype was recorded following Andersen et al. (2013).

Distribution

Rare in subfossil material, the only remain was recorded from a reservoir in the Štiavnické vrchy Mts. (SK).

Ecology and palaeoecological interpretation

Larvae of the only European species, *Paratrissocladius excerptus* (Walker, 1856), are rheophilic, live in sandy and muddy sediments of streams and rivers, prefer relatively fast-flowing medium size rives (metarhitral to metapotamal, Janecek & Moog, 2002); rare in both small (upper courses) and large rivers as well as in lakes (Moller Pillot 2013). Larvae build tubes using substrate particles. Records are usually from unpolluted waterbodies, and they seem not to tolerate organic pollution (Wilson & Ruse 2005).

Given that larvae are rheophilic, remains are very rare in lacustrine sediments (Brooks et al. 2007 do not list it). In case of occurrence, it may indicate strong inlet discharge.

Parorthocladius Thienemann (Fig. 5.20)

Head capsule brownish, triangular in shape. **Occipital margin** dark brown (Fig. 5.20a). **Antenna** 5-segmented. **Premandible** simple, with one apical tooth. Apical **mandibular** tooth shorter than combined width of 3 inner teeth (Fig. 5.20b). **Mentum** subtriangular with three equal sized median teeth and 4 pairs of lateral teeth along steeply sloping sides (Fig. 5.20b). **Ventromental**

FIGURE 5.19 *Paratrissocladius.* Detail of head capsule showing mentum

plates narrow with beard consisting of a tuft of long setae (often missing in the subfossil material) (Fig. 5.20b).

Remarks

The larval head of *Parorthocladius* and *Synorthocladius* (Fig. 5.25) are very similar, however *Synorthocladius* has two long median teeth on mentum (Fig. 5.25b), while *Parorthocladius* has three shorter teeth.

Morphotypes

One morphotype was identified without placement to species/type.

Distribution

Parorthocladius head capsules were rare, only found in three Tatra Mts. lakes (SK, PL).

Ecology and palaeoecological interpretation

Larvae are primarily rheophilic, inhabiting small, cold, flowing waterbodies, such as mountain brooks (Moller Pillot 2013), however, they can also occur in the littoral of oligotrophic lakes (Andersen et al. 2013). Only one species,

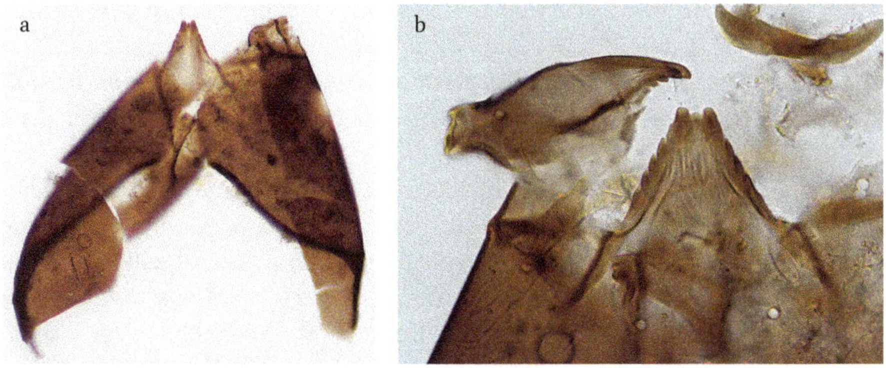

FIGURE 5.20 *Parorthocladius*. **a** head capsule, **b** mentum and mandible

P. nudipennis (Kieffer, 1908), is described as a larva out of the five known Palaearctic species (Andersen et al. 2013); the species is also known from the Tatra lakes (Bitušík 2004).

Rare in lake sediments; its presence may indicate strong inlet or extensive cold littoral zone.

Psectrocladius Kieffer (Fig. 5.21)
Head capsule yellow to light brown, **occipital margin** pale or darkened (Figs. 5.21a,c,f,j). **Antenna** 5-segmented, **premandible** with one tooth. Apical **mandibular** tooth shorter than or 2.5× longer than combined width of 3 inner teeth (Figs. 5.21e,i,l). **Mentum** with one or two median teeth and 5 pairs of subequal lateral teeth. When single median tooth, then either with one or two nipple-like median projections (Figs. 5.21b,d,g,k). **Ventromental plates** broad, extending beyond outermost mental teeth and forming a noticeable triangle at the apex (Fig. 5.21h). Beard consisting of variable number of setae present, but not always visible in subfossils.

Remarks
Psectrocladius is a relatively uniform genus, and subfossil head capsules provide a combination of characteristics that are distinctive: mentum with five lateral teeth and large acutely pointed ventromental plates with beard (which is often weakly visible or entirely missing in subfossil material), simple premandible (frequently missing in subfossils).

Morphotypes
Three types were distinguished following Brooks et al. (2007).

1 Mentum with a low, broad median tooth slightly taller than 1st lateral teeth 2
– Mentum with double median tooth or with a single broad tooth with 2 apical projections (in that case significantly taller than 1st lateral tooth) 3

2 Median mental tooth with distinct notches on the anterior margin, 2nd lateral tooth smaller than 1st and 3rd .... *Psectrocladius flavus*-type
– Median mental tooth with 2 low apical projections, lateral teeth subequal .... *Psectrocladius barbatipes*-type
3 Mentum with broad median tooth with 2 apical projections (weakly separated) .... *Psectrocladius psilopterus*-type
– Mentum with clearly separated 2 median teeth .... *Psectrocladius sordidellus*-type

***Psectrocladius* (*Allopsectrocladius*) *flavus*-type** (Fig. 5.21a). Mentum with one very broad median tooth with indentations (may appear as 3–4 median teeth); second lateral tooth clearly smaller than first and third (Fig. 5.21b).
***Psectrocladius* (*Mesopsectrocladius*) *barbatipes*-type** (Fig. 5.21c). Mentum with a very broad median tooth medially with a pair of short (often indistinct) projections; lateral teeth darker than median tooth and subeqeual in size (Fig. 5.21d).
***Psectrocladius* (*Psectrocladius*) *psilopterus*-type** (Fig. 5.21f). Mentum with single broad median tooth with pair of median nipple-like projections that are weakly separated. (Fig. 5.21g).
***Psectrocladius* (*Psectrocladius*) *sordidellus*-type** (Fig. 5.21j). Mentum with two, clearly separated median teeth (Fig. 5.21k). If only half of head capsule is preserved, it can be difficult to distinguish this morphotype from *P. psilopterus*-type.

Distribution
Psectrocladius remains were among the most common orthoclads in the lakes of all lake districts except the Maramures Mts. (RO). In the Tatra Mts. (SK, PL), remains were mostly found in lakes situated at lower elevations.

Ecology and palaeoecological interpretation
Larvae of this large genus are typically limnobionts, occupying virtually all types of stagnant waters, such as lakes (from oligotrophic to strongly eutrophic, Bitušík 2000), ponds, dams and peat bogs. The few records of the genus from flowing waters are often the results of dispersal from surrounding pools, dams and river arms (Moller Pillot 2013). The genus is characteristic for the littoral

zone (Lindegaard 1992, Moog 2002); some species are associated with macrophytes (Brodersen et al. 2001b). Species of the genus show different response to pH; while some species are associated with calcareous waters (e.g., *P. barbimanus*), some are acidotolerant to acidophilic (Moller Pillot 2013) and may become abundant in acidified lakes (Brooks et al. 2007 and references therein). The genus covers a wide range of temperature conditions: *P. barbatipes*-type and *P. psilopterus*-types are closer to the warm end of the gradient, while *P. sordidellus*-type being restricted to colder lakes (Brooks et al. 2007, Eggermont & Heiri 2012).

Subfossil remains are common in lake sediments. Their presence can be indicative of extensive littoral (with or without macrophytes) and acidic conditions.

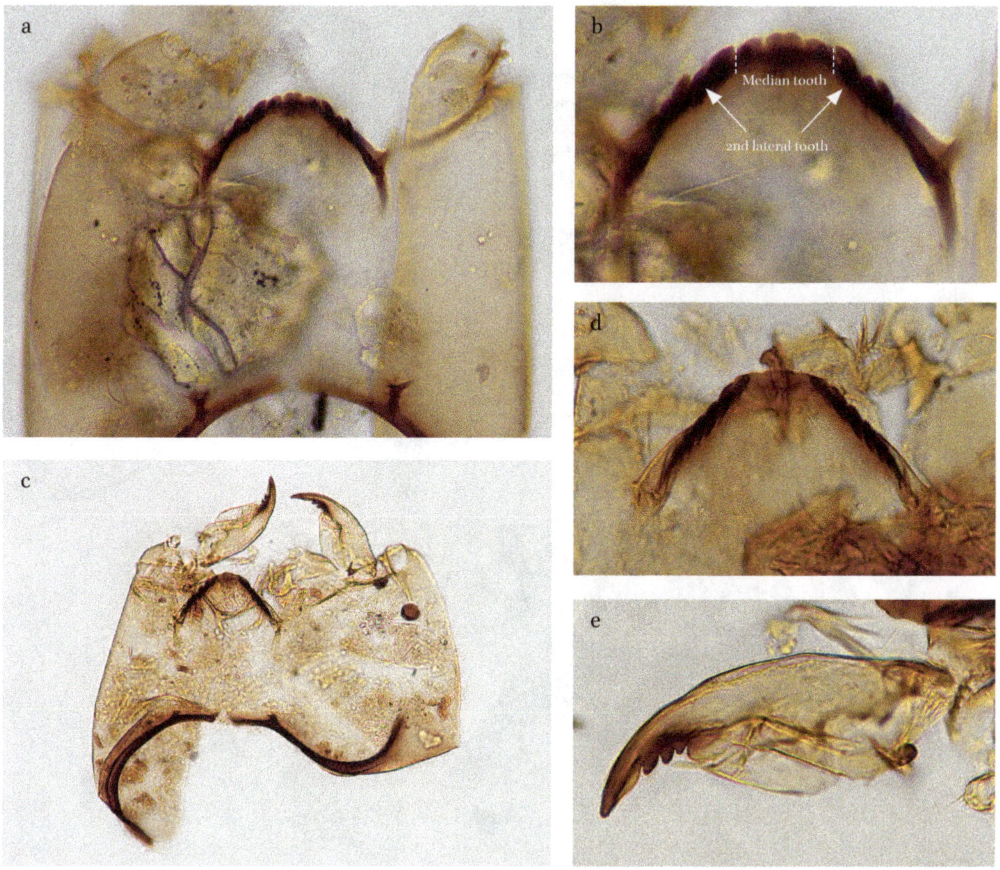

FIGURE 5.21 *Psectrocladius. Psectrocladius flavus*-type: **a** head capsule, **b** mentum. *Psectrocladius barbatipes*-type: **c** head capsule, **d** mentum, **e** mandible

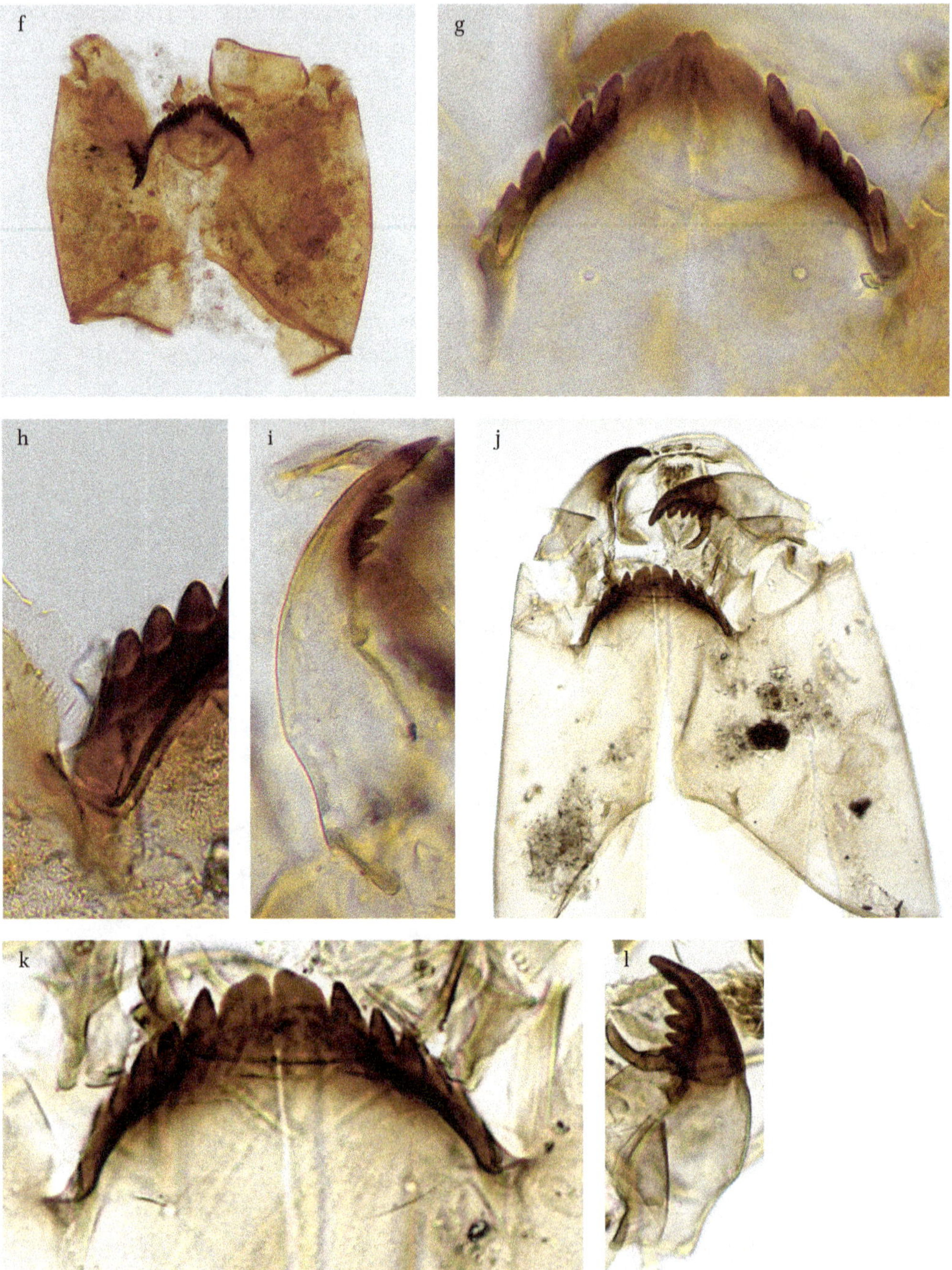

FIGURE 5.21 (continuation) *Psectrocladius psilopterus*-type: **f** head capsule, **g** mentum, **h** detail showing triangular ending of ventromental plate, **i** mandible. *Psectrocladius sordidellus*-type: **j** head capsule, **k** mentum, **l** mandible

Pseudosmittia Edwards (Figure 63) (Fig. 5.22)
Head capsule yellow, **occipital margin** pale to brown (Fig. 5.22a). **Antenna** reduced (only ¼ length of mandible), with 4 segments. **Premandible** with 2 apical teeth and 1–2 inner more or less tooth-like projections can be present. **Mandible** with 3–4 inner teeth (Fig. 5.22c). One very broad **median** mental tooth with a distinct median projection and 4 pairs of lateral teeth gradually decreasing in size (Fig. 5.22b). **Ventromental plates** large, not extending beyond outermost teeth (Fig. 5.22b).

Remarks
The shape of mentum is distinctive among other Orthocladiinae with exception of *Camptocladius* (not shown in this book), however its larvae are fully terrestrial, so their presence in lake sediments is very unlikely. *Pseudosmittia* could be confused with *Pseudorthocladius* (not shown in this book) when its median mental teeth are worn but setae submenti is situated further below outermost lateral teeth compared to that of *Pseudosmittia*.

This heterogeneous genus was divided into three genera: *Allocladius*, *Hydrosmittia* and *Pseudosmittia* (see Ferrington & Sæther 2011). Since their separation in subfossils is not possible, here we deal with *Pseudosmittia* sensu lato.

FIGURE 5.22 *Pseudosmittia*. **a** head capsule, **b** mentum, **c** mandible

Distribution

Even though the genus is relatively common in Holarctic lake sediments (Brooks et al. 2007), it was only recorded in a Tatra Mts. lake (PL).

Ecology and palaeoecological interpretation

Pseudosmittia is a large genus with poorly known larval stages. Most of the known species are semiterrestrial/semiaquatic, however, some species are truly aquatic and occupy littoral and splash zone of lakes (Cranston et al. 1983), and macrophytes (Brodin 1986).

In the palaeomaterial it may indicate extensive littoral and intensive erosive processes in the catchment.

Rheocricotopus Brundin (Fig. 5.23)

Head capsule mostly yellowish with brown **occipital margin** (Fig. 5.23a). **Antenna** 5-segmented, **premandible** single (but may appear bifid due to an apical notch). Apical **mandibular** tooth shorter than width of 3 inner teeth. **Mentum** with double median tooth (single tooth is exceptional) and 5 pairs of lateral teeth and with or without small accessory tooth between median and first lateral tooth (Figs. 5.23b,c). **Ventromental plates** strongly developed, extending beyond outermost laterals, broader in apical half; beard dense, consisting of numerous setae (Fig. 5.23b).

Remarks

If bearded ventromental plates are well preserved in subfossils, it is unlikely to confuse them with other Orthocladiinae. Heads without accessory tooth may superficially resemble *Psectrocladius sordidellus*-type (Fig. 5.21j), however, strongly developed ventromental plates with dense beard is distinctive for *Rheocricotopus*.

Morphotypes

Two morphotypes were identified following Schmid (1993) in Brooks et al. (2007).

1 Small accessory tooth present between median and 1st lateral tooth .... *Rheocricotopus fuscipes*-type

– Accessory tooth absent between median and 1st lateral tooth .... *Rheocricotopus effusus*-type

***Rheocricotopus effusus*-type**. Accessory teeth outside median tooth absents (Fig. 5.23b).

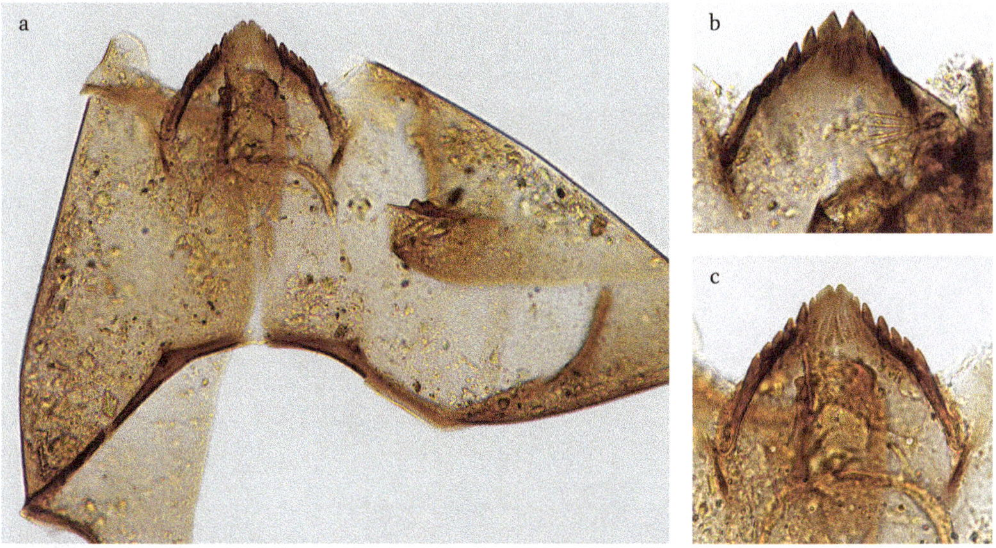

FIGURE 5.23 *Rheocricotopus*. **a** head capsule. *Rheocricotopus effusus*-type: **b** mentum. *Rheocricotopus fuscipes*-type: **c** mentum

***Rheocricotopus fuscipes*-type.** Accessory teeth outside median tooth present (Fig. 5.23c). Morphotype *chalybeatus* has similar shape of mentum but setae submenti are situated posterior to ventromental plates (Brooks et al. 2007).

Distribution

Rare in the subfossil dataset. Two morphotypes were recorded in the sediments of the reservoirs in the Štiavnické vrchy Mts. (SK) and one (*R. effusus*-type) in the Chornohora Massif (UA) with intense cold inlet.

Ecology and palaeoecological interpretation

Generally, larvae of *Rheocricotopus* are rheophilic and live on plants and hard anorganic substrata in rivers and streams. Rarely, larvae can be found in the splash zone of the littoral of oligotrophic lakes (Bitušík 2000).

Subfossil remains indicate enhanced inlet and/or increased littoral zone with coarse mineral sediments in cold lakes (e.g., Hamerlík et al. 2016).

Smittia Holmgren/*Parasmittia* Strenzke (Fig. 5.24)

Head capsule yellow to brownish, **occipital margin** pale or darkened (Fig. 5.24a). **Antenna** reduced, with 4 segments. Blade well developed and as long as flagellum (*Smittia*, Fig. 5.24c) or considerably longer (*Parasmittia*). **Premadible** with 2 apical teeth and 1 variably developed inner tooth. **Mandible** with 3 inner

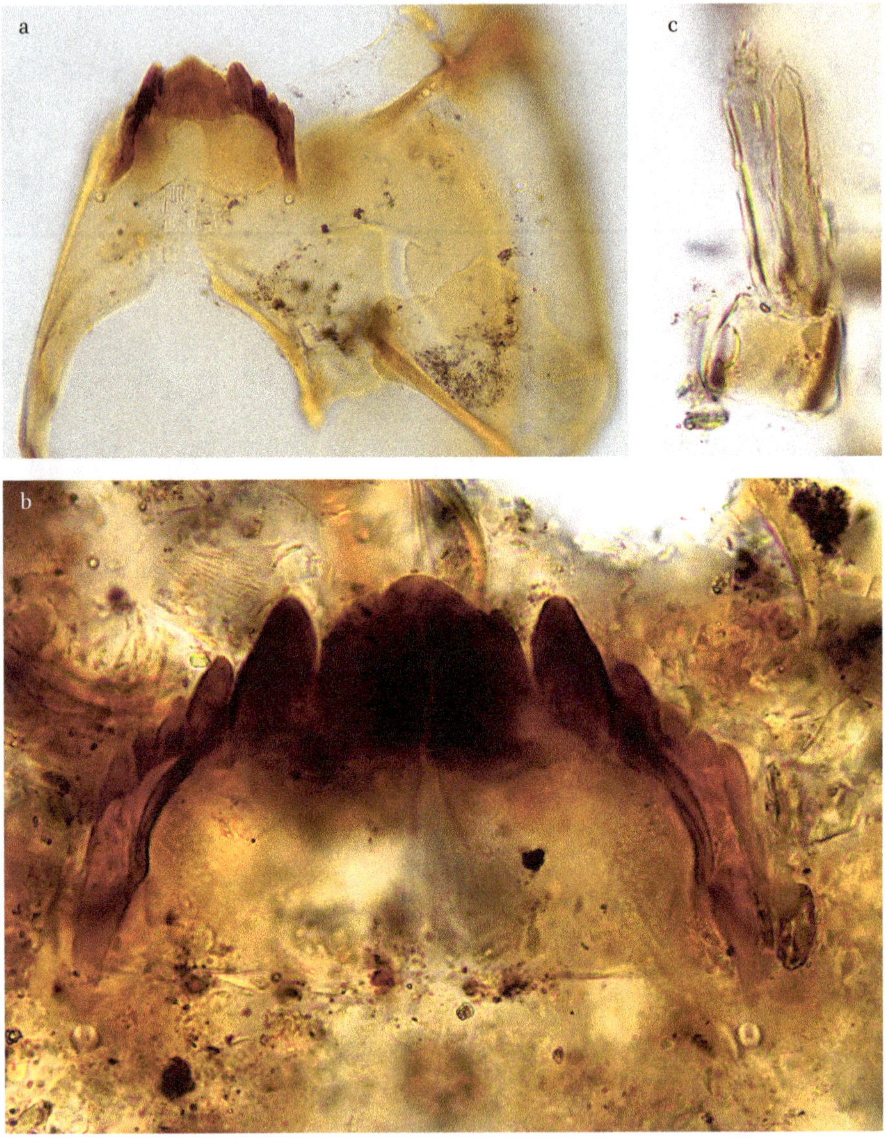

FIGURE 5.24 *Smittia/Parasmittia*. **a** head capsule, **b** mentum, **c** antenna (note the reduced 1st segment)

teeth and a short apical tooth. **Mentum** with one broad median tooth with projecting nipple (may be domed in *Smittia*) and 5 pairs of subequal lateral teeth (Fig. 5.24b). **Ventromental plates** well developed, not extending beyond outermost teeth.

Remarks

Smittia and *Parasmittia* are indistinguishable in specimens with missing antennae.

Distribution

Even though *Smittia/Parasmittia* occurred in all lake districts surveyed, their remains were relatively scarce and never abundant.

Ecology and palaeoecological interpretation

Most species of this probably large genus (probably, because the number of species is still unclear, and very few larvae are described; Andersen et al. 2013) are terrestrial living in damp soils. Few species are truly aquatic.

Smittia/Parasmittia head capsules are common in Palaearctic lake sediments (Brooks et al. 2007). Their remains may indicate water-level fluctuations, erosion events or changes in the ratio of lake margin lengths to lake area.

Synorthocladius Thienemann (Fig. 5.25)

Head capsule yellow, triangular (Fig. 5.25a). **Antenna** 5-segmented. **Premandible** with one apical tooth. **Mandible** with 3 inner teeth and short apical tooth, mola with one strong spine (Fig. 5.25c). **Mentum** triangular with two median teeth and 4 or 5 pairs of lateral teeth sloping steeply (Fig. 5.25b). **Ventromental plates** narrow and long with strong beard (may be missing in subfossil material).

Remarks

Larval *Synorthocladius* and *Parorthocladius* (Fig. 5.20) are similar, however mentum of *Synorhocladius* has 2 median teeth on its apex as opposed to *Parorthocladius* with 3 median teeth (Fig. 5.20b).

Morphotypes

One morphotype was identified, *Synorthocladius semivirens*-type following Andersen et al. (2013).

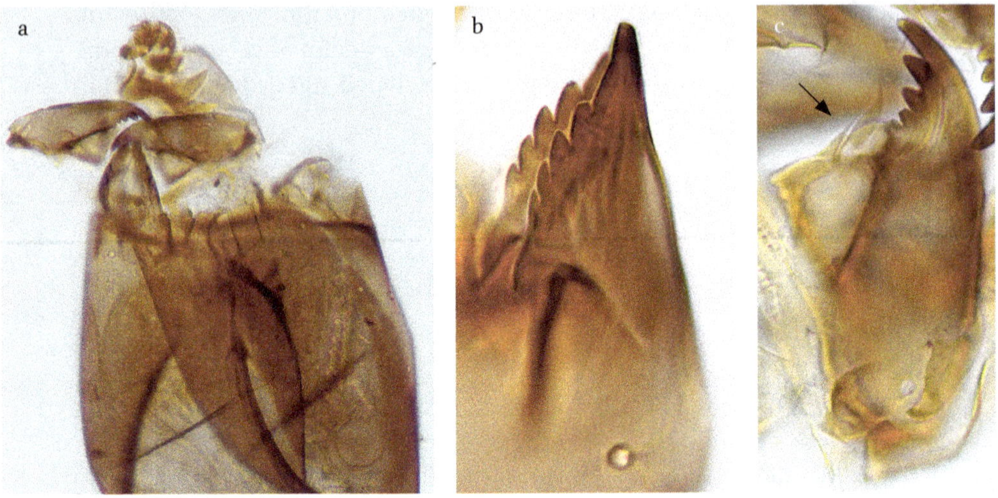

FIGURE 5.25 *Synorthocladius*. **a** head capsule, **b** mentum, **c** mandible (arrow indicates spine)

Distribution

Subfossil remains were rare, only found in two Tatra Mts. lakes (SK, PL) and two reservoirs in the Štiavnické vrchy Mts. (SK).

Ecology and palaeoecological interpretation

The only Holarctic species, *Synorthocladius semivirens* (Kieffer, 1909) belongs to the most common chironomid species in Slovakia (Bitušík 2000). Larvae are eurytopic and eurythermic, inhabiting springs, various bodies of flowing water from rhitral to potamal as well as small waterbodies and littoral zone of lakes (Moog 2002, Andersen et al. 2013).

Remains of *Synorthocladius* are usually neither frequent nor abundant in lake sediment samples (Brooks et al. 2007). Due to their eurytopic character, it is difficult to define their value as palaeoindicators, but perhaps they could indicate extensive littoral or strong inlet.

Thienemanniella Kieffer (Fig. 5.26)

Head capsule yellow to brown, **occipital margin** dark brown (Fig. 5.26a). Compared with other orthoclads (with exception of *Corynoneura*), head capsule is prolonged in shape and very small, easy to overlook. **Antenna** at least 1/2 but no more than 3/4 length of head; 5-segmented; segment 3 is frequently darkened. **Premandible** with one apical tooth. **Mandible** with 4 inner teeth, first one can be longer than apical tooth (Fig. 5.26b). **Mentum** with three equal median teeth clearly above the 5 pairs of subequal lateral teeth (Fig. 5.26b).

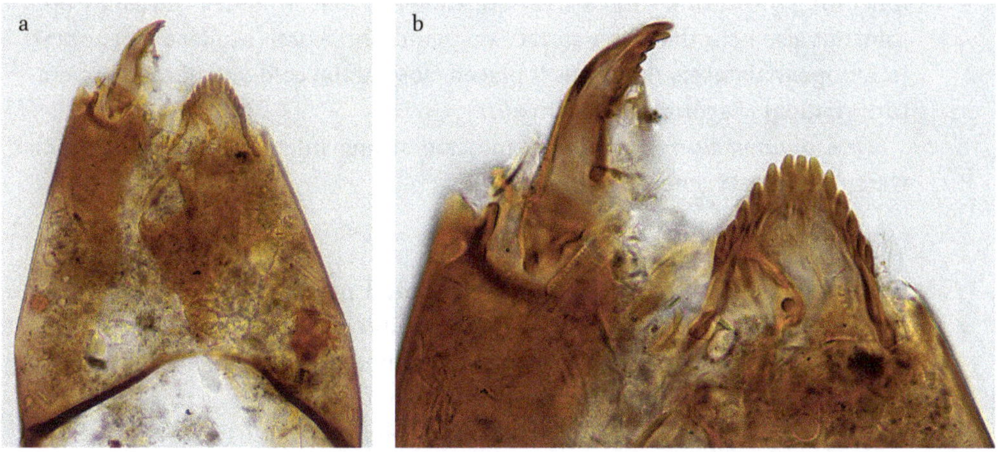

FIGURE 5.26 *Thienemanniella clavicornis*-type: **a** head capsule, **b** mentum and mandible

Remarks

Thienemaniella resembles *Corynoneura* (Fig. 5.4) and differs from it in the number of antennal segments, which is 5 as opposed to 4 in *Corynoneura*. As antennae are often missing in subfossils, usually darker head capsule always without net-like reticulation on the surface will differentiate *Thienemanniella* from *Corynoneura*.

Morphotypes

One morphotype was identified following Brooks et al. (2007).

***Thienemanniella clavicornis*-type** (Fig. 5.26a). Central median tooth is approximately the same length as the outer pair of median teeth, first inner tooth of mandible as long as apical tooth (Fig. 5.26b).

Distribution

It is important to note that due to their small size, *Thienemanniella* head capsules (and also larvae) are easy to overlook and as a consequence, the genus is under-collected and understudied. Records in lake sediments most likely do not reflect the real distribution of the genus. Head capsule remains were found in low abundance in several Tatra Mts. lakes (SK, PL) located at lower altitudes and in one lake in the Chornohora Massif (UA).

Ecology and palaeoecological interpretation

Larvae are primarily rheophilic occupying various types of lotic habitats, from fast flowing mountain streams to upper sections of lowland rivers (eupotamal) (Moog 2002). Larvae are also found in lakes (Moller Pillot 2013), most likely inhabiting the littoral zone. Larvae are known to live on stones and sandy bottom, but also near the water surface on plants and wood (Moller Pillot 20013). In European datasets the genus is placed close to the cold end of the temperature gradient (Eggermont & Heiri 2012).

Thienamanniella remains may indicate strong inlet of cold, oxygen rich water or well-oxygenated littoral zone.

Tvetenia Kieffer (Fig. 5.27)
Head capsule yellow to light-brown, **occipital margin** pale or dark brown (Fig. 5.27a). **Antenna** 5-segmented, **premandible** with one apical blunt tooth. Apical tooth of **mandible** shorter than combined width of 3 inner teeth, mola usually with 2–3 spines, some species without spines (Fig. 5.27c). **Mentum** striped, with 1 or 2 median teeth and 5 pairs of lateral teeth (Fig. 5.27b). **Ventromental plates** narrow creating an appearance of dark bulge below last lateral tooth of mentum (Fig. 5.27b). **Setae submenti** is located well below mentum (Figs. 5.27a,b).

Remarks
Tvetenia closely resembles *Eukiefferiella* (Fig. 5.8), but the combination of yellow color of head with paired median teeth and position of setae submenti well below mentum are distinctive for *Tvetenia*.

Morphotypes
One morphotype was identified following Brooks et al. (2007).
***Tvetenia bavarica*-type.** Head capsule yellow with pale occipital margin (Fig. 5.27a). Mentum with two median teeth and 5 pairs of lateral teeth (Fig. 5.27b).

Distribution
Subfossil remains were scarce, recorded in some Tatra Mts. lakes (SK, PL) and in all studied reservoirs in the Štiavnické vrchy Mts. (SK).

Ecology and palaeoecological interpretation
Larvae are primarily rheophilic and inhabit various types of flowing waters from eucrenal to metapotamal (Moog 2001); they can also occur in the littoral of oligotrophic lakes (Bitušík 2000, Krno et al. 2006). Larvae prefer hard substrates, such as gravel and stones, but also live among mosses and algae growing on hard substrates (Moller Pillot 2013 and references therein).

Remains indicate increased inlet influence (Bitušík et al. 2018) and/or well oxygenated littoral with coarse sediments (Hamerlík et al. 2016).

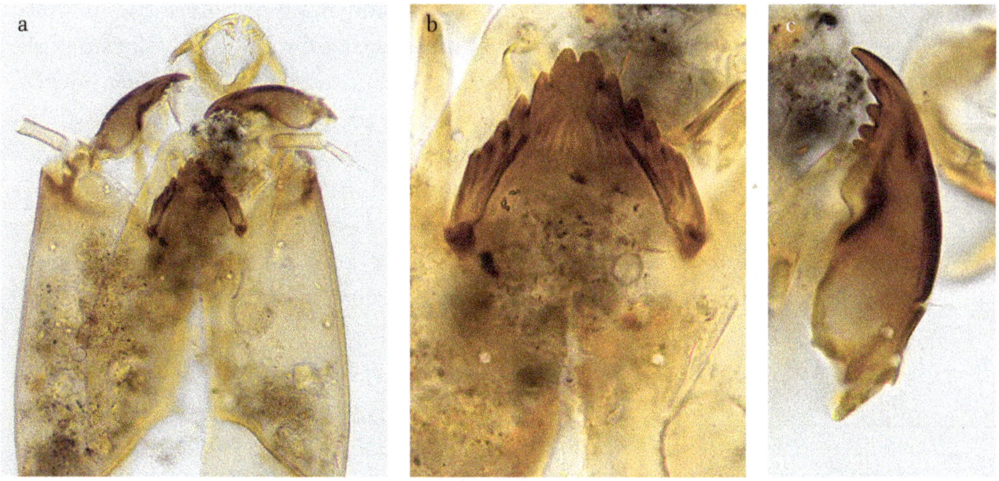

FIGURE 5.27 *Tvetenia bavarica*-type: a head capsule, **b** mentum and position of seta submenti, c mandible

Zalutschia Lipina (Fig. 5.28)

Head capsule yellow with brown **occipital margin** (Figs. 5.28a,f). **Antenna** with 6 segments, sixth segment vestigial (Figs. 5.28e,i). **Premandible** bifid apically with one inner tooth. Apical **mandibular** tooth shorter than combined width of 3 inner teeth (Figs. 5.28d,g,h). **Mentum** usually dark with double median tooth and 6 pairs of lateral teeth, first lateral reduced, occasionally with 2–3 light colored median teeth and first and sixth lateral teeth reduced (Figs. 5.28b,c,g). **Ventromental plates** well developed extending beyond mentum with fine beard (Figs. 5.28b,g).

Remarks

Zalutschia is a distinctive genus. It resembles *Hydrobaenus* (Fig. 5.11) which has similar shape of mentum, however, beard is not present. *Zalutschia* could also be confused with *Psectrocladius* (Fig. 5.21) due to distinct ventromental plates with beard, however mentum of *Psectrocladius* has five lateral teeth only and first lateral teeth are not reduced.

Morphotypes

Two morphotypes were identified following Brooks et al. (2007).

1 Apical tooth of mandible and median teeth of mentum lighter than occipital margin; outermost mental tooth minute, mandible with 3rd inner tooth subequal in size to other inner teeth .... *Zalutschia mucronata*-type

- Mandible and mentum darker than occipital margin; outermost mental tooth not significantly reduced, mandible with 3rd tooth markedly smaller than remaining inner teeth .... *Zalutschia* type B

***Zalutschia mucronata*-type.** Head capsule light brown, occipital margin dark brown (Fig. 5.28a). Teeth of mandible and mentum lighter than occipital margin. Outermost mental tooth strongly reduced in size (Figs. 5.28b,c). Inner mandibular teeth decreasing in size, third inner tooth not significantly smaller than second tooth (Fig. 5.28d). Basal antennal segment shorter than flagellum, blade longer than flagellum (Fig. 5.28de).
***Zalutschia* type B.** Head capsule yellow, occipital margin light brown (Fig. 5.28f). Teeth of mandible and mentum strongly pigmented, darker than occipital margin. Outermost mental tooth not significantly reduced (Fig. 5.28g). Mandible with first and second teeth subequal in size, third tooth markedly smaller (Fig. 5.28h). Basal antennal segment longer than flagellum (length of blade unknown, but larvae with long basal antennal segment have blade at most as long as flagellum, see Sæther 1976) (Fig. 5.28i).

Distribution
Subfossil *Zalutschia* remains were recorded in a small number (5) of mostly acidified Tatra Mts. lakes (SK, PL).

Ecology and palaeoecological interpretation
Larvae of *Zalutschia* are typical inhabitants of the littoral and profundal of oligotrophic to mesotrophic lakes (Moller Pillot 2013) and tend to dominate in polyhumic and acidic alpine lakes (Bitušík et al. 2006, Krno et al. 2006) and ponds (Hamerlík et al. 2017). Larvae were recorded among *Sphagnum* and filamentous algae as well as on organic bottoms (Moller Pillot 2013).

Appearance or increase of the proportion of *Zalutschia* remains in the palaeorecord will be most likely connected to the decrease of pH.

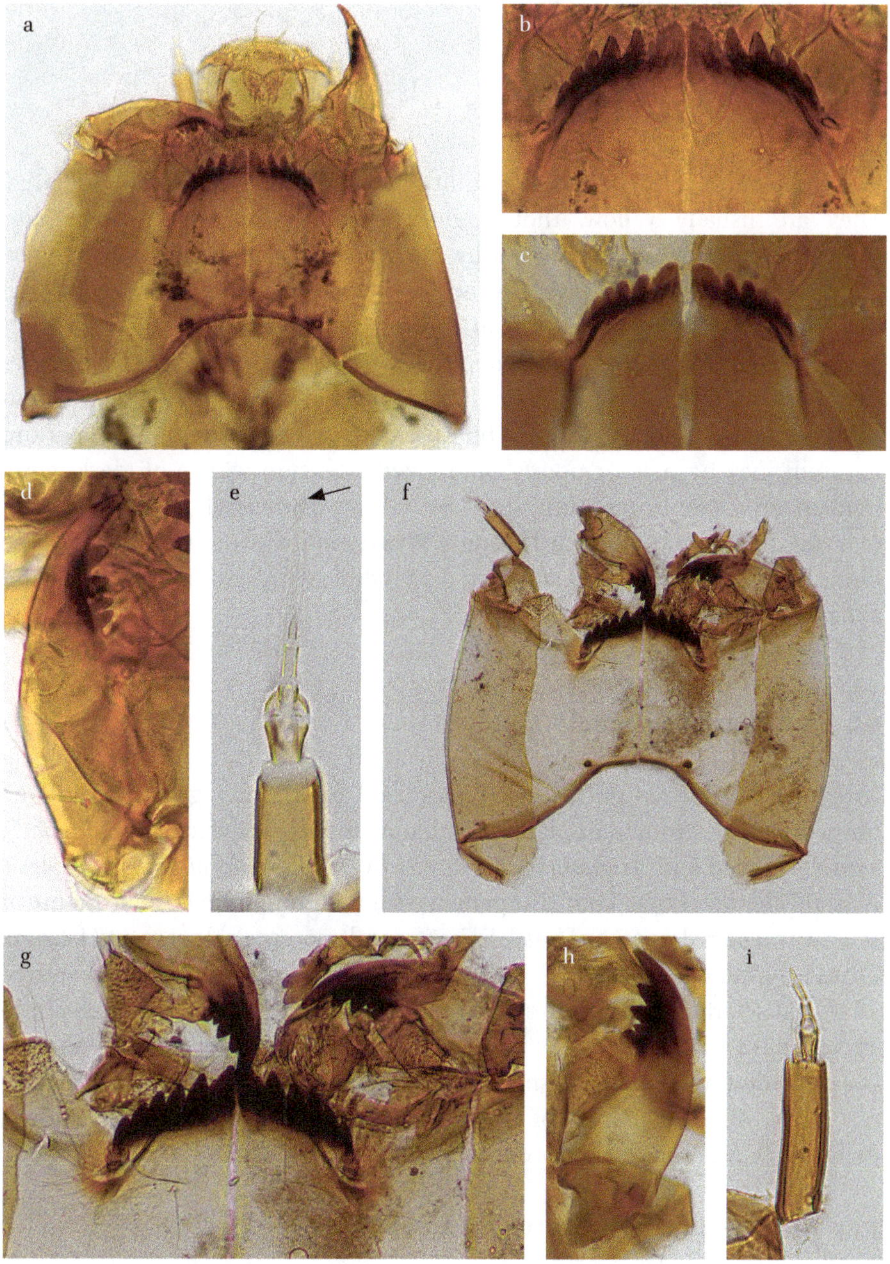

FIGURE 5.28 *Zalutschia*. *Zalutschia mucronata*-type: **a** head capsule, **b** mentum, **c** alternative shape of mentum with worn down teeth, **d** mandible, **e** antenna (arrow indicates end of blade). *Zalutschia* type B: **f** head capsule, **g** mentum, **h** mandible, **i** antenna

CHAPTER 6

Chironominae

Head capsules of larval Chironominae vary in size from small to large. They are usually yellow, though brown patches may be present dorsally. The submentum may be fully darkened, or a darker patch may be restricted to the area above the occipital margin. Antennae always well developed, 5–8 segmented. They arise from rounded protuberances (Chironomini, Pseudochironomini) or are placed on distinct pedestals (Tanytarsini), frequently with distal spur. Premandibles well developed, with 1–7 teeth. Mentum usually strongly sclerotized, but sometimes median teeth or the entire mentum may be pale or poorly sclerotized. Mentum with 9–16 dark teeth, median area (ventromentum) may be visibly delineated from the lateral regions (dorsomentum). Beard absents. Ventromental plates variable in shape but usually well developed and striated. Mandible with several inner teeth and usually with dorsal tooth.

6.1 Tribe Chironomini

Antennae not placed on distinct pedestals, Lauterborn organs not situated on pedicels. Ventromental plates variable in shape (but usually fan-shaped), well-developed and striated (reduced only in the *Stenochironomus* complex). A useful character to separate the genera within Chironomini is the structure of sclerites on dorsal surface of head: frons (frontal apotome), clypeus and labrum. Frons is either discrete or fused with clypeus and forms frontoclypeus (rarely, all dorsal sclerites fused to form frontoclypeolabrum – *Stenochironomus*). In some genera, there is a distinct fenestra (Fig. 6.7e) at anterior frons or frontoclypeus, sometimes reduced to smaller mark, line or area with different structure of cuticle. Anterior margin of frons or fronclypeus can be straight, curved, expanded laterally, sometimes crenulated, or with tubercles on surface.

KEY TO GENERA OF THE TRIBE CHIRONOMINI

1 Mentum with single median tooth 2
– Mentum with 2 or more median teeth 7

 | DOI:10.1163/9789004694781_007

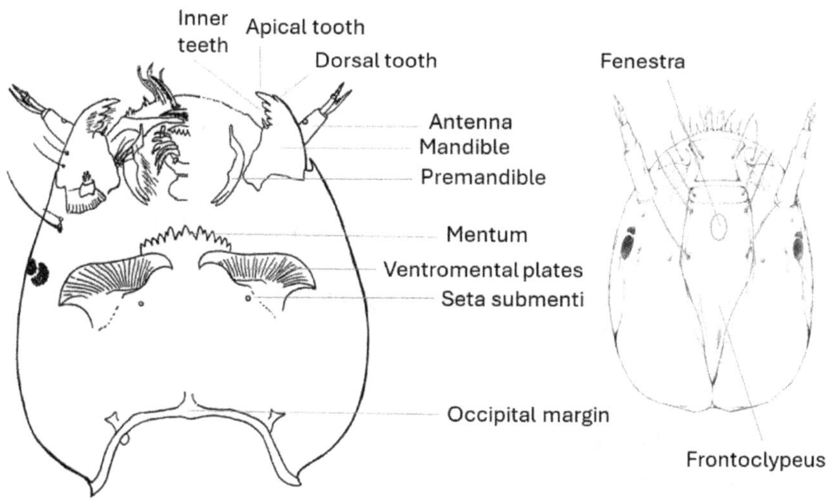

FIGURE 6.1 Chironomini larval head capsule showing ventral head (left) comparing diagnostic features of a recent (left side) and subfossil remain (right side), and dorsal head (right). Ventral head: modified after Bitušík (2000), dorsal head: drawing by Nera Vuić

2 Mentum with wide median tooth (3 and more times the width of 1st laterals), domed and weakly pigmented 3
– Mentum with narrow median tooth, up to 2× as broad as 1st lateral tooth, normally pigmented 4

3 Mentum concave, outermost lateral tooth notched apically .... *Cryptochironomus*
– Mentum convex, 3 outermost lateral teeth arranged in distinct cluster .... *Cryptotendipes*

4 Ventromental plates narrow, fan shaped, about half width of mentum .... *Dicrotendipes*
– Ventromental plates broad 5

5 Mentum with median tooth clearly taller than 1st lateral tooth ..... *Benthalia*
– Mentum with median tooth subequal to 1st lateral tooth 6

6 Ventromental plates triangular, short (up to the width of mentum), anterior margin coarsely scalloped .... *Parachironomus*
– Ventromental plates prolonged, at least as long as width of mentum, anterior margin smooth or finely crenulated .... *Glyptotendipes*

7 Mentum with 2 median teeth 8
– Mentum with 3 or more median teeth 16

8 Mentum with 6 pairs of lateral teeth 9
– Mentum with 7 pairs of lateral teeth 11

9 Outermost 3 lateral teeth of mentum forming a distinct cluster, 6th lateral tooth taller than adjacent teeth .... *Cladopelma* in part (*C. laccophila*-type)
– Outermost lateral teeth not forming a distinct cluster, subequal in size ... 10

10 Ventromental plates triangular with straight apical margin, 1st and 2nd lateral teeth of mentum separate .... *Lauterborniella*
– Ventromental plates curved, 1st and 2nd lateral teeth partly fused .... *Microtendipes*

11 Outermost 3 lateral teeth of mentum forming a distinct cluster, 6th lateral tooth taller than adjacent teeth .... *Cladopelma* in part (*C. lateralis*-type)
– Outermost lateral teeth not forming a distinct cluster, subequal in size ... 12

12 Mentum with 1st lateral tooth clearly reduced in size (shorter or narrower) than 2nd lateral tooth 13
– Mentum with 1st and 2nd lateral teeth subequal 14

13 1st lateral tooth narrow, needle-like, ventromental plates strongly curved .... *Pagastiella*
– 1st lateral tooth short, at most half of the height 2nd lateral tooth, ventromental plates slightly curved .... *Polypedilum* in part

14 Each median tooth about as broad as 2nd lateral tooth .... *Polypedilum* in part
– Median teeth broader, > 2× as broad as 2nd lateral teeth 15

15 Mentum convex, ventromental plates long, about same width as mentum, coarsely striated .... *Paracladopelma*
– Mentum horizontal, ventromental plates short, about half the width of mentum, finely striated .... *Harnischia*

16 Mentum with 3 median teeth 17
– Mentum with 4 median teeth 19

17 Central median tooth of mentum minute .... *Microtendipes* (*M. pedellus*-type)
– Central median tooth of mentum taller or subequal to outer median teeth 18

18 Outer median teeth reduced in size, about half the size of median tooth .... *Chironomus*
– Outer median teeth longer or about the same size as median tooth .... *Endochironomus* in part (*E. tendens*-type)

19 Ventromental plates narrow and elongate, lateral apex curved into a lobe .... *Endochironomus* in part (*E. impar*-type, *E. albipennis*-type)
– Ventromental plates short, lateral apex truncate 20

20 Mentum with 1st lateral tooth clearly shorter than 2nd lateral tooth 21
– Mentum with 1st lateral tooth at least as tall as 2nd lateral tooth 22

21 1st lateral tooth shorter than outer median tooth, mentum uniformly colored, strongly pigmented .... *Phaenopsectra*
– 1st lateral tooth higher than outer median tooth, median mental teeth of similar height, weakly pigmented, pale, outer laterals dark ... *Paratendipes*
– 1st lateral tooth subequal to outer median tooth, central pair of median teeth sunken and slender, outer pair broad and slope inward, mentum uniformly pigmented .... *Omisus*

22 Mandible with 4 inner teeth, anterior margin of ventromental plates with short paired striation .... *Sergentia*
– Mandible with 2–3 inner teeth, anterior margin of ventromental plates with broad grey band .... *Stictochironomus*

Note that some genera where number of median mental teeth is unclear may be keyed twice (e.g., *Microtendipes pedellus*-type with both 2 and 3 medial teeth).

Benthalia Lipina (Fig. 6.2)
Head capsule large, yellow with brown submentum, **occipital margin** black (Fig. 6.2a). **Antenna** 5-segmented (Fig. 6.2b), **premandible** with 2 slender apical teeth and 2 short basal teeth. **Mandible** with two pale, indistinct dorsal teeth, 2–3 inner teeth and radially arranged grooves on ventral surface of mandible base (Fig. 6.2d). **Mentum** with single median tooth widening to

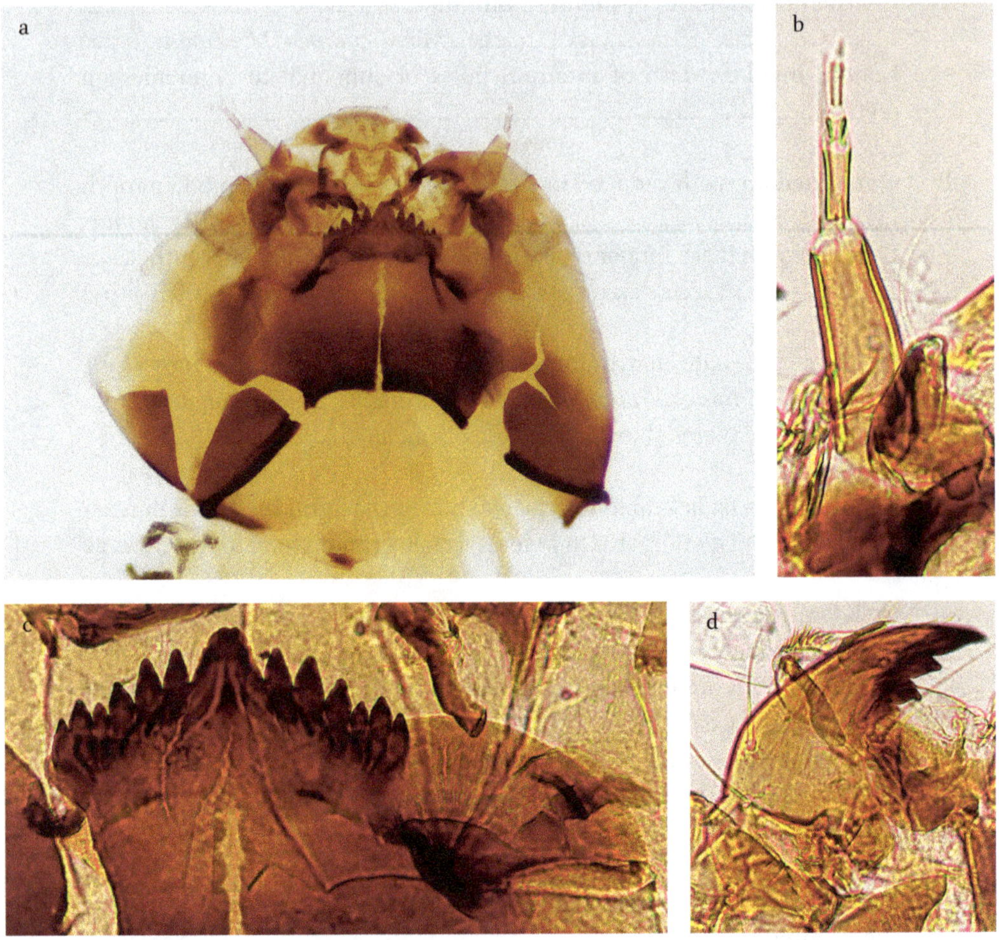

FIGURE 6.2 *Benthalia*. **a** head capsule, **b** antenna, **c** mentum, **d** mandible

base, lateral notch very short or absent and with 6 pairs lateral teeth, fourth lateral shorter than fifth (Fig. 6.2c). **Ventromental plate** broad, anterior margin smooth, striations faint (Fig. 6.2c).

Remarks

The generic name *Benthalia* was adopted for species previously belonging to *Einfeldia* group B and D (Pinder & Reiss 1983) when morphological characteristics (namely shape of pecten epipharyngis) were supported by molecular data (Martin et al. 2007). *Benthalia* subfossil remains could be confused with *Fleuria* (currently subgenus of *Chironomus*) having taller and more robust median mental tooth than any lateral teeth. Median tooth in *Fleuria* (not shown in this book) is only slightly widening to its base and lateral notch is distinct in unworn conditions. Moreover, in *Fleuria* a field of short spinules is present between labral lamellae, however, this character is hardly applicable to subfossil material.

Morphotypes

The morphotype identified corresponds to *Einfeldia natchitocheae*-type with two mandibular inner teeth listed in Brooks et al. (2007, Fig. 6.17).

Distribution

Subfossil material was only collected from sediments of Morské oko lake (SK).

Ecology and palaeoecological interpretation

Larvae inhabit mainly organic bottom without vegetation in the littoral of lakes and ponds, rarely can be found in slow flowing waters. They are common in eutrophic and somewhat polluted conditions with low oxygen content (see Moller Pillot 2009 and citation therein).

Subfossil remains are indicative of shallow part of eutrophic lakes with soft sediments outside of macrophyte cover.

Chironomus Meigen (Fig. 6.3)

Head capsule yellow, often with brown pigmented submentum and/or dorsal part of head (frontoclypeus). **Occipital margin** dark brown to black (Figs. 6.3a,d). **Antenna** 5-segmented. **Premadible** typically with 2 teeth (very rarely more, 3–5). **Mandible** usually with 2 pale dorsal teeth (often hardy visible), dark apical tooth, and 2–3 inner teeth; basally on ventral surface radially arranged grooves (Figs. 6.3c,f). **Mentum** with trifid median tooth (*Fleuria*, recently considered as a subgenus of *Chironomus* has simple median tooth with short lateral incision), 6 pairs of lateral teeth, first 2 mostly closely appressed (Figs. 6.3b,e). **Ventromental plates** broad, usually each plate about as wide as mentum or somewhat wider (Figs. 6.3b,e).

Remarks

The shape of the three median teeth is distinctive among other Chironomini except for *Einfeldia pagana*-type (Brook et al. 2007; not shown in this book), which has a large heart-shape fenestra on frontoclypeus. As mentioned above, subgenus *Fleuria* (not shown in this book) is very similar to *Benthalia* by the shape of the mentum but can be distinguished using other characters (see remarks to *Benthalia*).

Morphotypes

Two morphotypes were recognised following Brooks et al. (2007).

1 Mentum with 4th lateral tooth smaller than 3rd and 5th .... *Chironomus anthracinus*-type
– Mentum with lateral teeth decreasing gradually in size, 4th lateral tooth longer than 5th .... *Chironomus plumosus*-type

***Chironomus anthracinus*-type.** Fourth lateral tooth of mentum shorter than adjacent lateral teeth (Fig. 6.3b). Mandible with two inner teeth (Fig. 6.3c).
***Chironomus plumosus*-type.** Lateral teeth of mentum decreasing gradually in size, fourth lateral tooth longer than fifth (Fig. 6.3e). Mandible with three inner teeth (Fig. 6.3f).

In our material, however, there are also individuals combining the *plumosus*-type mentum with *anthracinus*-type mandible and the other way around. Considering the great amount of *Chironomus* species (more than hundred) with different ecological requirements more studies focusing on differentiation of further morphotypes is needed.

Distribution

Chironomus subfossils were found in mesotrophic lakes in the Tatra Mts. (SK, PL), the Chornohora and Svydovets Massifs (UA), the Maramures Mts. (RO), Morské oko lake and in all studied reservoirs in Štiavnické vrchy Mts. (SK).

Ecology and palaeoecological interpretation

Chironomus is a species rich genus with great variability in ecological requirements. Generally, larvae are typical bottom dwellers, living in tubes on substrata with high content of organic material. They prefer stagnant waters of different size and depth (including temporary water bodies), and can be found in slowly flowing rivers, often organically polluted. Larvae of many species can survive in waters with low oxygen content, some are even hypoxia resistant; larvae of other species inhabit acid conditions with very low pH (see Moller Pillot 2009 and citation therein). In the Tatra Mts. training-set, both types are located on the warm end of the temperature gradient with *C. plumosus*-type having the highest July air-temperature optimum (11.9 °C), while that of *C. anthracinus*-type is slightly lower, 11.4 °C.

As larval densities can occasionally reach up to tens of thousands of individuals per m^2, *Chironomus* remains are often numerous in subfossil records and indicate usually warm, productive conditions. Increase of *Chironomus* throughout the sediment record indicates a transition of a lake into higher trophic state (eutrophication) due to increased nutrient inputs connected to oxygen depletion at the lake bottom (Brodersen & Quinlan 2006 and references therein). We found short-lived dominance of *Chironomus anthracinus*-type

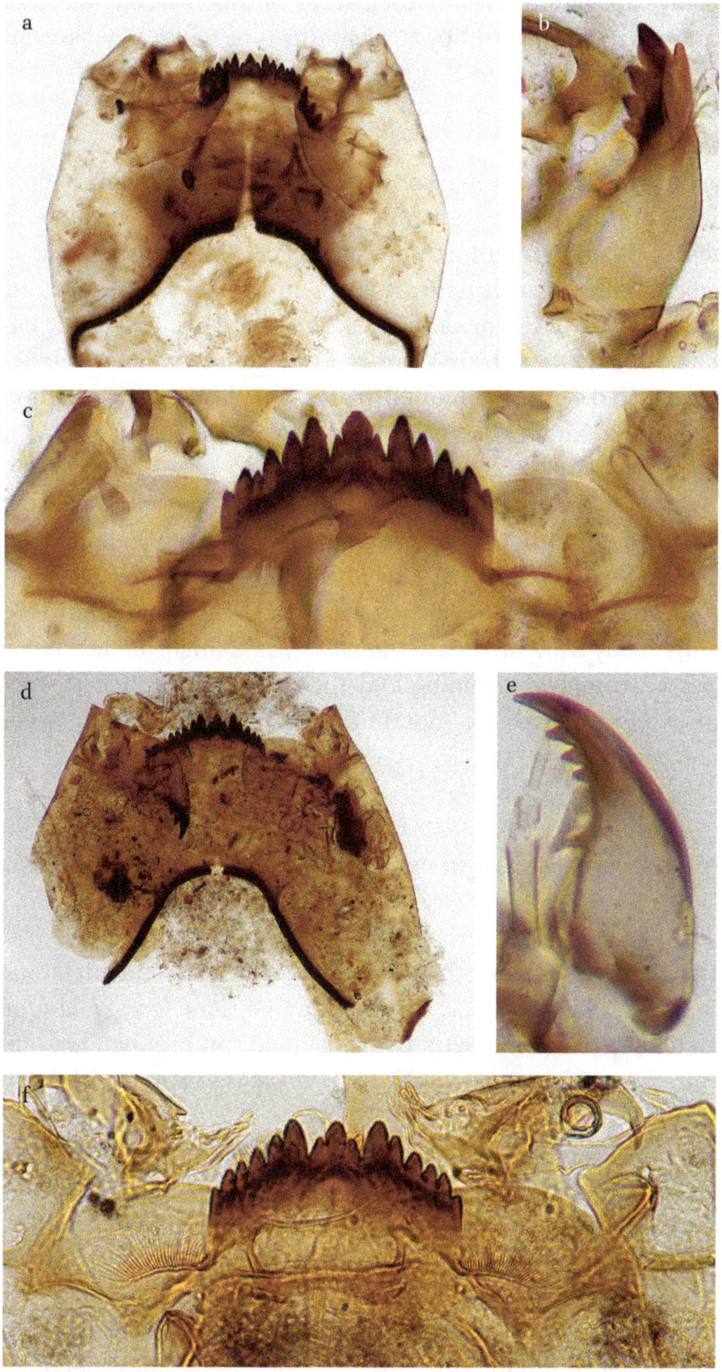

FIGURE 6.3 *Chironomus*. *Chironomus anthracinus*-type: **a** head capsule, **b** mandible, **c** mentum. *Chironomus plumosus*-type: **d** head capsule, **e** mandible, **f** mentum

during unstable hydrology (significant oscillation of water-level) of Veľká Richňava reservoir that was replaced by *C. plumosus*-type when the reservoir was permanently flooded (Bitušík et al. 2018).

Cladopelma Kieffer (Fig. 6.4)
Head capsule light, greyish with brownish submentum, **occipital margin** dark brown (Fig. 6.4a). **Antenna** with 5 segments, premandible bifid. Dorsal tooth on **mandible** absent, 1–2 flat inner teeth present (Fig. 6.4c). **Mentum** usually with double median tooth (or single with a median notch) and 7 pairs of lateral teeth. Outermost lateral teeth set forward in relation to slope of remaining lateral teeth, 5th and 7th lateral teeth are short and the 6th lateral tooth is taller and broader (Figs. 6.4b,d,e). **Ventromental plates** broad, with fine striation distinct at base, anterior margin often crenulated (Fig. 6.4b).

Remarks
Mentum with double median tooth and a separated group of outer lateral teeth is distinctive for *Cladopelma* (Fig. 6.4b). *Cryptotendipes* and *Microchironomus* (not shown in this book) have a similar pattern of the lateral teeth, however median tooth is distinctly trifid in *Microchironomus* while single median tooth in *Cryptotendipes* is strongly protruding and the mentum is much steeper (Fig. 6.6b) than in *Cladopelma*; the mentum of *Microchironomus* is rather horizontal.

Morphotypes
Two morphotypes were recognized in our material corresponding to Brooks et al. (2007).

1 Mentum with 6 pairs of lateral teeth, broad gap between 3rd and 4th lateral teeth .... *Cladopelma laccophila*-type
– Mentum with 7 pairs of lateral teeth, without broad gap between 3rd and 4th lateral teeth .... *Cladopelma lateralis*-type

***Cladopelma laccophila*-type.** Three outermost lateral teeth separated from the remaining teeth by a deep notch, forming a separate group (Fig. 6.4e). The morphotype corresponds to *C. viridulum* group according to Moller Pillot (2009) and Vallenduuk (2019).
***Cladopelma lateralis*-type** (Figs. 6.4a–d). Group of outermost lateral teeth not widely separated from the remaining laterals (Figs. 6.4d). This morphotype corresponds to *C. goetghebueri* group sensu Moller Pillot (2009) and Vallenduuk (2019).

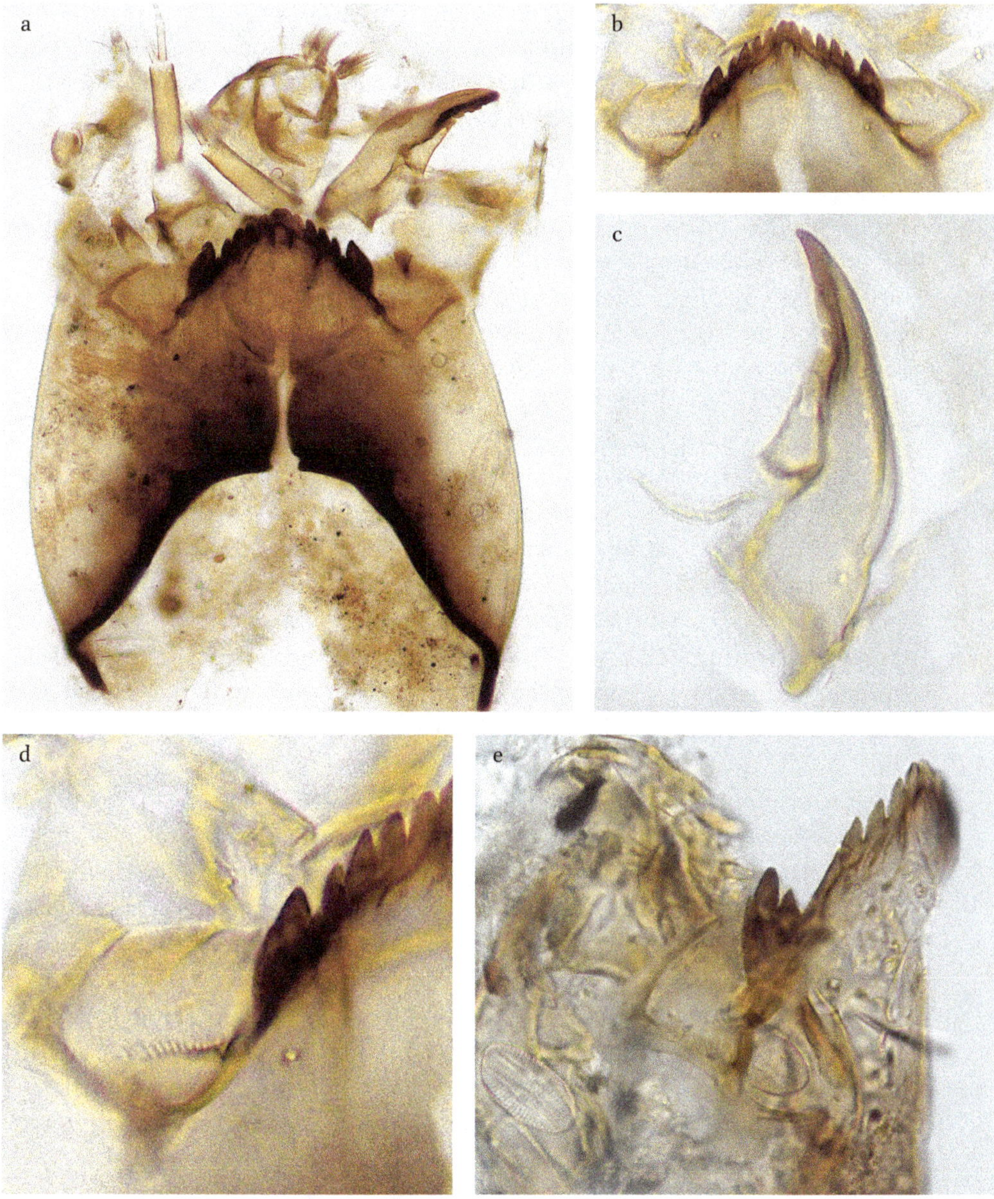

FIGURE 6.4 *Cladopelma. Cladopelma lateralis*-type: **a** head capsule, **b** mentum, **c** mandible; **d,e** comparison of menta of *C. lateralis*-type (**d**) and *C. laccophila*-type (**e**)

Distribution

Remains of both types were found in a reservoir in Štiavnické vrchy Mts. (SK), in two dystrophic lakes situated in the forest zone of the Tatra Mts. (SK, PL) and in Morské oko lake (SK) (*Cladopelma lateralis*-type).

Ecology and palaeoecological interpretation

Larvae of *Cladopelma* are bottom dwellers, living mostly in the littoral and sublittoral (rarely in profundal) of lakes, in smaller stagnant waters and in brooks and large rivers. Larvae of species belonging to *viridulum* group (i.e. *laccophila*-type) are often found in acid waters and they are tolerant to eutrophic conditions with low oxygen contents (Rossaro et al. 2007). According to Moller Pillot (2009) larvae of the group are more often collected in eutrophic environments than *goetghebueri* group (*C. lateralis*-type), whose larvae have been collected from standing waters ranging from oligotrophy to eutrophy.

The decline of the relative abundance of *Cladopelma* remains in sediment records is usually connected with increased lake productivity (Brodersen & Quinlan 2006). Generally, *Cladopelma* is considered warm stenothermal and confined to rather mesotrophic lakes (Brooks et al. 2007 and references therein, Samartin et al. 2012). However, different morphotypes/groups differ in ecological requirements and thus for precise interpretation they should be distinguished.

Cryptochironomus Kieffer (Fig. 6.5)

Head capsule dark yellow (orange in live larvae), posteriorly swollen (visible in lateral projection), submentum sometimes with obvious pigmentation, **occipital margin** usually pale (Fig. 6.5a). **Antenna** 5-segmented, **premandible** with 4–6 teeth declining in size towards base. **Mandible** with long apical tooth and 2 triangular inner teeth, dorsal tooth absent (Fig. 6.5c). Median tooth of **mentum** pale and broad, 6 pairs of lateral teeth arranged obliquely (outermost lateral tooth higher than first lateral tooth); first lateral tooth usually fused with the median tooth and the outermost tooth may be notched (Fig. 6.5b). **Ventromental plates** large, wider than mentum and tapering to point (Fig. 6.5b).

Remarks

Cryptochironomus resembles *Demicryptochironomus* (not shown in this book) in shape of mentum, but the latter has the first (of seven) lateral tooth separated from the median tooth, and the outermost lateral tooth is without an apical notch.

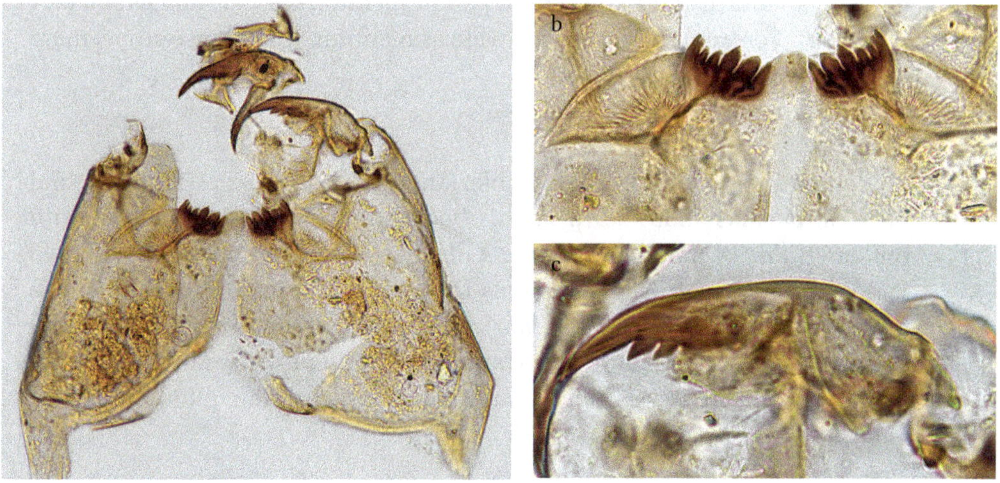

FIGURE 6.5 *Cryptochironomus*. **a** head capsule, **b** mentum, **c** mandible

Distribution

Subfossil remains were found frequently in small numbers in the sediments of all surveyed reservoirs in the Štiavnické vrchy Mts. (SK).

Ecology and palaeoecological interpretation

Cryptochironomus larvae occur in various types of flowing and stagnant waters regardless of trophy or saprobity level, they do, however, require sufficient concentration of dissolved oxygen (Moller Pillot 2013 and references therein). Larvae can ventilate oxygen-rich water from above the sediment and migrate to more oxygen-saturated habitats when the deeper parts of the lake bottom become hypoxic (Bazzanti et al. 1998, Brodersen & Quinlan 2006).

Subfossil remains in sediment records can indicate well-oxygenated littoral or profundal, sandy substrata with some detritus and fine organic material but without dense aquatic vegetation, and sufficient amount of food, i.e. oligochaetes, and small chironomid larvae (Valenduuk & Morozova 2005).

Cryptotendipes Beck & Beck (Fig. 6.6)

Head capsule yellow-brown, submentum often darker than remaining head, **occipital margin** brown (Fig. 6.6a). **Antenna** 5-segmented. **Premadible** with 2 slender teeth apically. Apical **mandibular** tooth short, 2 wide and flat inner teeth, dorsal tooth absent. **Mentum** with broad, rounded median tooth (sometimes notched laterally, giving trifid appearance) and 6 pairs of lateral teeth; outermost 3 teeth forming a distinct cluster; fifth pair is obviously more robust than fourth and sixth; central part of mentum steeper than lateral part (Fig. 6.6b). **Ventromental plates** as wide as mentum, with coarse striae, more distinct posteriorly (Fig. 6.6b).

Remarks

Mentum of *Cryptotendipes* resembles that of *Cladopelma* (Fig. 6.4) and *Microchironomus* (not shown in this book), having the same pattern with the outermost lateral teeth forming a distinct group and mandible with two flattened inner teeth (see comments to *Cladopelma*). In addition to simple median tooth (double in *Cladoplema*), it differs from both genera in the overall shape of mentum: the slope of the anterior margin of the central part of the mentum is much steeper and gradually levels off with the outer lateral part being horizontal.

Distribution

Subfossil remains were common but never abundant in the sediments of two reservoirs in the Štiavnické vrchy Mts. (SK).

Ecology and palaeoecological interpretation

The genus is thermophilous in general, so larvae absent from lakes above 1000 m a.s.l. (Reiss 1968). *Cryptotendipes* larvae have been recorded in lakes from the littoral down to 11 m of depth, as well as in slow flowing waters. Larvae inhabit sand, silt and coarse decaying plant material but avoid places with anoxia (Moller Pillot 2013 and references therein).

Cryptotendipes remains are usually found in low densities in sediment records and are indicative of meso- to eutrophic conditions.

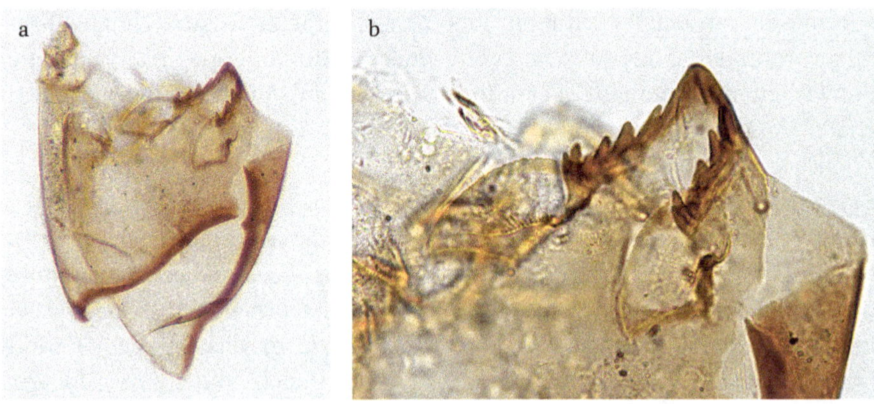

FIGURE 6.6 *Cryptotendipes*. **a** head capsule, **b** mentum

Dicrotendipes Kieffer (Fig. 6.7)
Head capsule yellow, **occipital margin** dark (Figs. 6.7a,e). **Frontal apotome** separated from clypeus (except for one species) (Figs. 6.7d,e), and there is an antero-median fenestra (Fig. 6.7e) or mark (Fig. 6.7d) on the frontal apotome; anterior margin of frontal apotome may be crenulated (Fig. 6.7d). **Antenna** 5-segmented, segment 4 exceptionally long. **Premandible** with 3 teeth, second one and third broad and blunt. **Mandible** with pale dorsal tooth, 1–2 small surficial and 3 inner teeth (Figs. 6.7b,g). Median tooth of mentum robust, sometimes laterally notched, 6 pairs of lateral teeth. First and second pairs sometimes partially fused. Sixth lateral tooth sometimes reduced into a broad lobe, so only five laterals are clearly visible (Figs. 6.7b,g). **Ventromental plates** narrow, visibly fan-shaped, completely striated, about half width of mentum; anterior margin smooth (Fig. 6.7f) or crenate (Fig. 6.7c).

Remarks
Mentum of *Dicrotendipes* reminds of those of *Glyptotendipes* (Fig. 6.9), *Einfeldia* and *Kiefferulus* (the latter two are not shown in this book), moreover, all have (at least some of the species) a fenestra on frontal apotome. They can be separated by the shape and size of ventromental plates that are short and distinctly fan-like in *Dicrotendipes* while long (equalling to or exceeding width of mentum) in the other three mentioned genera.

Morphotypes
Two morphotypes were distinguished following Brooks et al. (2007).

1 Anterior margin of ventromental plates crenulated; mentum rather triangular in shape, outermost lateral tooth reduced .... *Dicrotendipes nervosus*-type
– Anterior margin of ventromental plates smooth; mentum rather flat, outermost lateral teeth separate .... *Dicrotendipes notatus*-type*... 2

2* Anterior margin of frontal apotome straight to slightly concave (Fig. 6.7e), smooth or at most slightly crenate, fenestra oval, situated close to the anteriour margin of frontal apotome .... *Dicrotendipes notatus*
– Anterior margin of frontal apotome markedly convex (Fig. 6.7i) with strong tuberculation in front of and around the circular fenestra which is located further from the anteriour margin of frontal apotome .... *Dicrotendipes lobiger*

* In case the median part of the dorsal head (i.e., dorsal sclerites) is visible, it is possible to distinguish two species of the morphotype following Contreras-Lichtenberg (1984).

***Dicrotendipes nervosus*-type** (Fig. 6.7a). Anterior margin of ventromental plates crenulated (Fig. 6.7c). Anterolateral margin of frontal apotome coarsely crenulated with median mark, antero-median fenestra missing (Fig. 6.7d). Mentum rather triangular (compared to that of *D. notatus*-type).
***Dicrotendipes notatus*-type** (Fig. 6.7e). Ventromental plates with smooth anterior margin (Fig. 6.7f). Oval or circular antero-median fenestra present on frontal apotome (Fig. 6.7e). Within this type two species can be distinguished if dorsal sclerites are present. *Dicrotendipes notatus* (Meigen, 1818) has the anterior margin of frontal apotome straight to slightly concave (Fig. 6.7e), smooth or at most slightly crenate, fenestra oval, relatively small, situated close to the anterior margin of frontal apotome. The characteristic feature of *Dicrotendipes lobiger* (Kieffer, 1921) is the markedly convex anterior margin of frontal apotome (Figs. 6.7h,i) and the strong tuberculation in front of and around the fenestra. The fenestra is circular and larger than in *D. notatus* and also located further from the anterior margin of frontal apotome.

Distribution

Subfossil remains of *Dicrotendipes nervosus*-type and *D. notatus*-type were found in Morské oko lake, reservoirs in the Štiavnické vrchy Mts. (SK) and in some Tatra Mts. lakes (PL) with higher trophic state caused by direct anthropogenic impact (tourism). *Dicrotendipes lobiger* was only recorded in Vrbické pleso lake, Low Tatra Mts. (SK).

Ecology and palaeoecological interpretation

Dicrotendipes is a typical dweller of various types of stagnant waters, although it can also be found in slow flowing streams and rivers. Larvae of some species live predominantly on plants (*D. nervosus*, *D. lobiger*), others prefer bottom substrata, both organic (e.g., leaves, wood) and inorganic, such as stones and artificial materials (*D. notatus* and *tritomus* group). *Dicrotendipes* is common in the littoral of oligotrophic to strongly eutrophic lakes; larvae of some species (*D. nervosus*, *D. notatus*) can survive low oxygen concentrations (< 1 mg L^{-1}) for a short time, while species from the *tritomus* group require well oxygenated conditions (Rossaro et al. 2007, Moller Pillot 2013 and references therein). In the Tatra Mts. training-set, the July air-temperature optimum of *T. notatus*-type is 9.9 °C.

Dicrotendipes remains will indicate shallow, oligotrophic to eutrophic lakes, possibly with rich aquatic vegetation (Brodersen & Quinlan 2006, Langdon et al 2010).

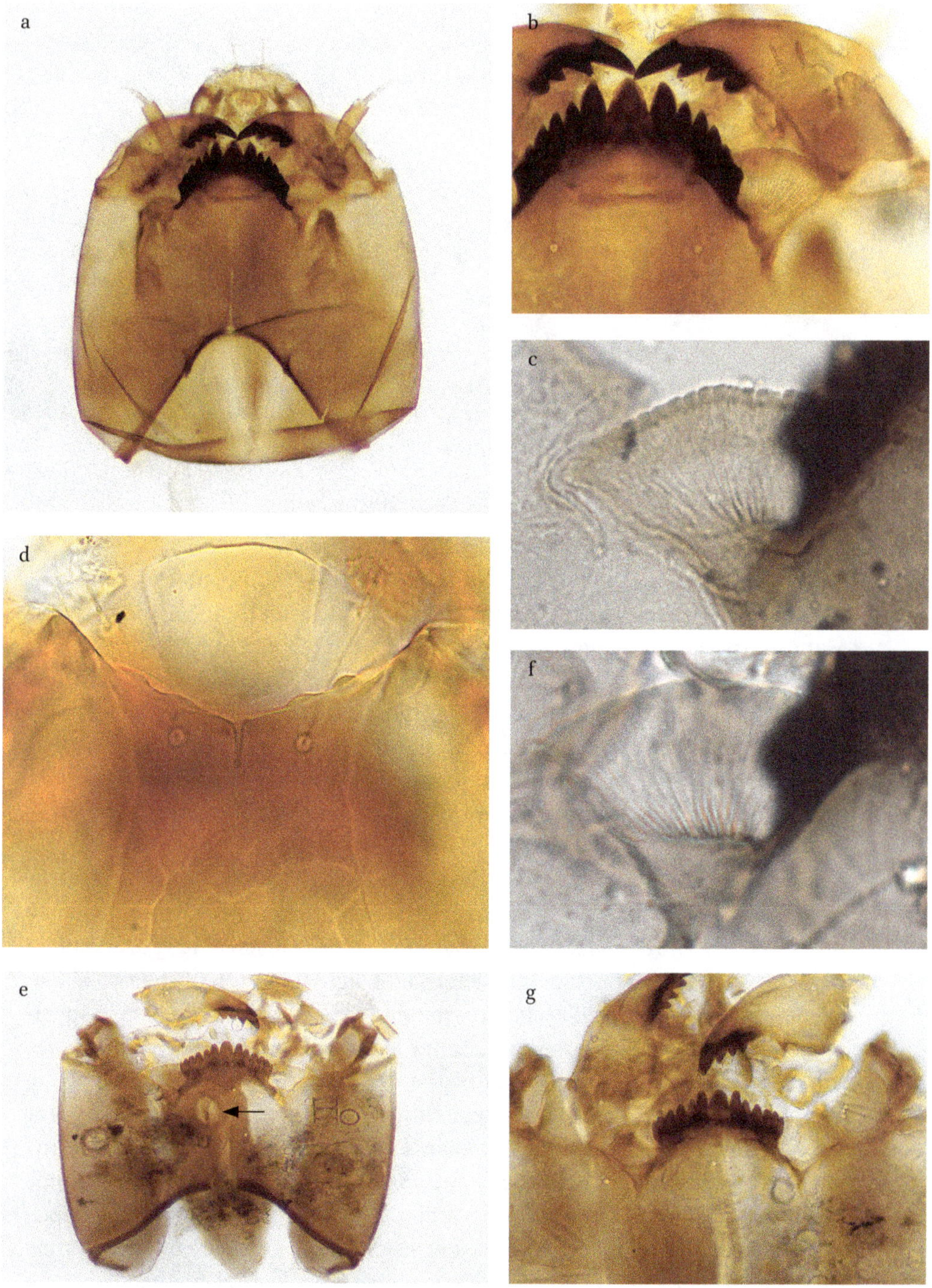

FIGURE 6.7 *Dicrotendipes*. *Dicrotendipes nervosus*-type: **a** head capsule, **b** mentum, mandible and ventromental plates, **c** detail of anteriour margin of ventromental plate, **d** anterolateral margin of frontal apotome. *Dicrotendipes notatus*-type: **e** head capsule (arrow indicates oval antero-median fenestra on frontal apotome), **f** detail of ventromental plate, **g** mentum, mandible and ventromental plates

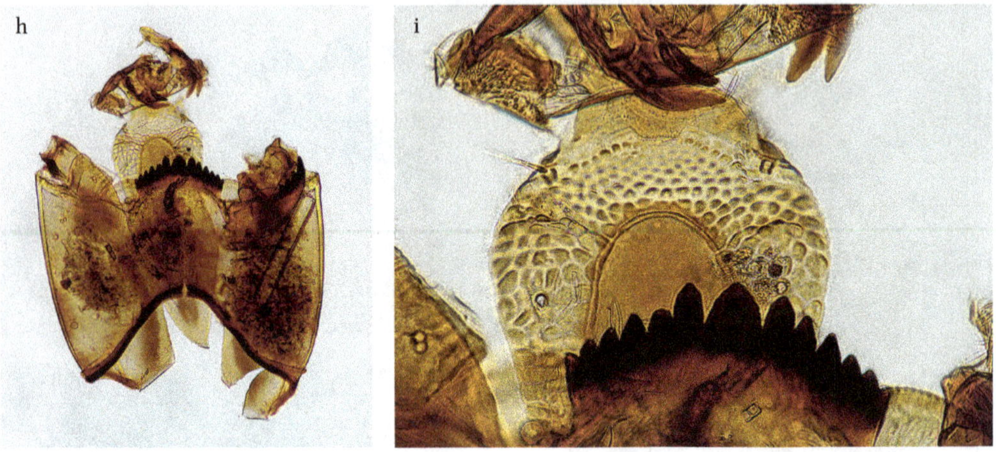

FIGURE 6.7 (continuation) *Dicrotendipes lobiger*-type: **h** head capsule, **i** anterior margin of frontal apotome

Endochironomus Kieffer/***Synendotendipes*** Grodhaus (Fig. 6.8)
Head capsule large, yellow, **occipital margin** dark brown (Figs. 6.8a,c). **Antenna** 5-segmented, **premandible** with 2 apical teeth and 1 inner tooth. **Mandible** with slender apical tooth and 3–4 inner teeth (3 in *Endochironomus*, 4 in *Synendotendipes*), dorsal tooth pale, hardly visible (Fig. 6.8e). **Mentum** with 3–4 protruding median teeth (central ones lower than outer pair) and 6 pairs of lateral teeth; first lateral tooth shorter than second lateral and median tooth (Figs. 6.8b,d,f). **Ventromental plates** relatively narrow and gently curved with smooth anterior margin, striation continuous (Figs. 6.8b,d,f).

Remarks
Genus *Synendotendipes* was erected based on immature stages previously placed in *Endochironomus* (= species group B, Pinder & Reiss, 1983). As subfossil heads may lack important characters allowing separation of the two above genera (mandibula, maxilla), it is reasonable to keep them together. In specimens where mandible is present, 4 inner teeth will separate *Synendotendipes* from *Endochironomus* with 3 inner teeth.

In the absence of other characters, subfossils can be confused with genera having similar menta with four median teeth and clearly delineated ventromentum, i.e. *Phaenopsectra, Sergentia, Stictochironomus* and *Tribelos* (not shown in this book). Ventromental plates of *Endochironomus/Synendotendipes* are longer, curved and lobed at the outer margin, while they are relatively truncate in the other similar taxa mentioned above (Brooks et al. 2007).

Morphotypes

Three morphotypes were found following Brooks et al. (2007).

1 Mentum with 3 median teeth .... *Endochironomus tendens*-type
– Mentum with 4 median teeth 2

2 Outer median teeth of mentum longer than inner teeth, 1st lateral tooth clearly shorter than 2nd lateral tooth; mandible with 4 inner teeth .... *Endochironomus impar*-type
– All 4 median teeth of mentum subequal in height, inner median teeth partly fused, 1st and 2nd lateral teeth subequal; mandible with 3 inner teeth .... *Endochironomus albipennis*-type

***Endochironomus albipennis*-type** (Fig. 6.8a). Four median mental teeth subequal in height, distinctly taller and more robust than lateral teeth, inner median teeth partly fused centrally. First lateral teeth subequal to second laterals (Fig. 6.8b). Mandible with 3 inner teeth.
***Endochironomus impar*-type** (Fig. 6.8c). Mentum with four median mental teeth, outer median teeth longer than inner ones, first laterals clearly shorter than second lateral teeth (Fig. 6.8d); mandible with 4 distinct inner teeth (Fig. 6.8e). This morphotype corresponds to genus *Synendotendipes* sensu Epler et al. (2013).
***Endochironomus tendens*-type.** Three median mental teeth, central tooth slightly shorter than outer median teeth. First lateral teeth shorter than second laterals (Fig. 6.8f). Mandible with 3 inner teeth.

Distribution

Endochironomus impar-type was common in acidic lakes in the Tatra Mts. (SK, PL) and in the Chornohora and Svydovets Massifs (UA). Subfossil remains of *Endochironomus albipennis*-type and *E. tendens*-type were found in reservoirs in the Štiavnické vrchy Mts. (SK).

Ecology and palaeoecological interpretation

Generally, larvae of *Endochironomus*/*Synendotendipes* species are closely associated with macrophytes in the littoral of stagnant waters; rarely can be found in slowly flowing streams. Larvae are usually leaf miners but can build nets also on the surface of leaves and decomposing organic material on the bottom. Occasionally they can be found on inorganic substrata (silt, sand, stones) as documented in an alpine pond without macrophytes (Novikmec

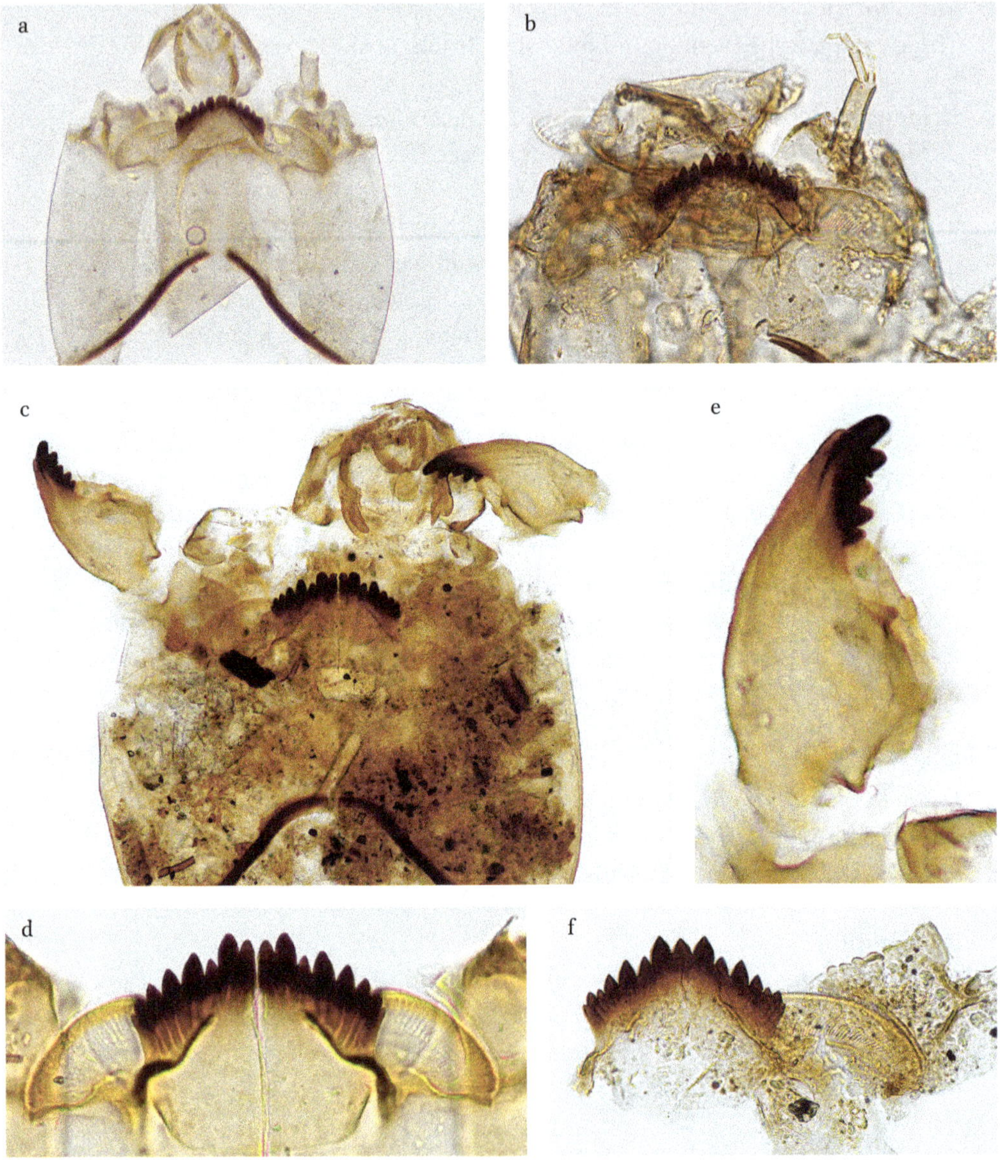

FIGURE 6.8 *Endochironomus*. *Endochironomus albipennis*-type: **a** head capsule, **b** mentum. *Endochironomus impar*-type: **c** head capsule, **d** mentum, **e** mandible. *Endochironomus tendens*-type: **f** mentum and ventromental plate

et al. 2015). Most *Endochironomus/Synendotendipes* species have been collected regularly in acidic water bodies (Brodin 1986, Moller Pillot 2013 and references therein). This corresponds well with the known distribution of *Endochironomus/Synendotendipes* in the acidic lakes of the Tatra Mts. and in the Chornohora and Svydovets Massifs, UA (Bitušík et al. 2006 and unpublished data).

Subfossil remains of *Endochironomus/Synendotendipes* may be indicative of macrophyte-covered littorals of mesotrophic to eutrophic lakes (e.g., Brodersen et al. 2001b, Brodersen & Quinlan 2006). Increase in the proportion of head capsules towards the present can be associated with anthropogenic acidification (Brodin 1986).

Glyptotendipes Kieffer (Fig. 6.9)

Head capsule yellow, **occipital margin** pale (Fig. 6.9a). **Frontal apotome** with or without long oval fenestra, anterior margin straight, convex or concave. **Antenna** with 5 segments. **Premandible** apically bifid with obvious inner tooth. **Mandible** with 3 inner teeth, but in some species basal inner tooth is only indicated, so appears as if only 2 inner teeth were present; dorsal tooth mostly present (Fig. 6.9b). **Mentum** with simple median tooth, sometimes laterally crenate; 6 pairs of lateral teeth, fourth sometimes shorter than third and fifth. **Ventromental plates** long and curved, medially separated by about 0.5–1.5 the width of median mental tooth or almost in contact; dorsal margin smooth or finely toothed, striae continuous (Fig. 6.9b).

Remarks

The shape of mentum resembles *Dicrotendipes* (Fig. 6.7) but the ventromental plates of *Glyptotendipes* are much longer and the first lateral teeth of mentum are parallel to the median tooth rather than oblique outwards as in *Dicrotendipes* (Fig. 6.9b). The length of ventromental plates will separate larval *Einfeldia* (not shown in this book) from species of *Glyptotendipes* with frontal fenestra (subgenus *Heynotendipes*).

Morphotypes

One morphotype was recognized following Brooks et al. (2007).

***Glyptotendipes pallens*-type.** Median tooth of mentum about the same height as the first lateral teeth, ventromental plates dorsally crenulated (Fig. 6.9b).

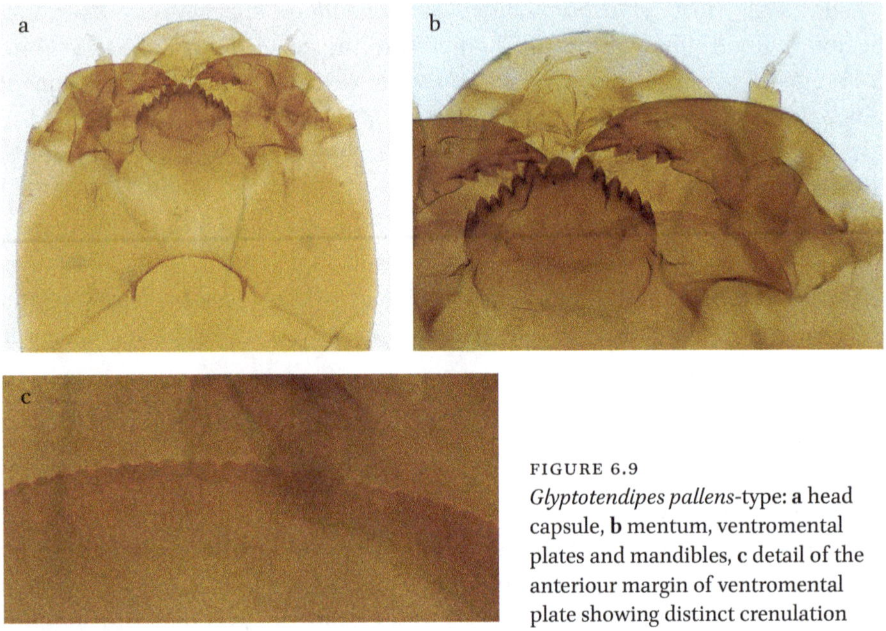

FIGURE 6.9
Glyptotendipes pallens-type: **a** head capsule, **b** mentum, ventromental plates and mandibles, **c** detail of the anteriour margin of ventromental plate showing distinct crenulation

Distribution

The morphotype was recorded only in the reservoirs in the Štiavnické vrchy Mts. (SK), where it was relatively common.

Ecology and palaeoecological interpretation

Larvae of *Glyptotendipes* occur predominantly in the littoral zone of warm meso- to eutrophic lakes, reservoirs and pools, rarely in very slow (sometimes polluted) flowing waters; some species were recorded in temporary waters, and some can survive in frozen sediments. They live both on bottom, firm substrates (stones, wood) or decaying and living plants; larvae construct tubes or mine in living and dead plant tissues and decaying wood (often using abandoned mines after the original inhabitants). Exceptionally they can live in bryozoan colonies (*G.* (*Heynotendipes*) *signatus* (Kieffer, 1909)). Despite some contradictory findings (see Moller Pillot 2009 and references therein), important prediction of larval occurrence is sufficient oxygen content.

In the sediment record, *Glyptotendipes* remains may by indicative of warm, aerobic, non-acidified conditions, higher trophic status and structured littoral with macrophytes, wood and decaying material (Brodersen et al. 2001b). The subgenus *Heynotendipes* (not shown in this book, for illustration see Epler et al. 2013) is recognizable by a long oval fenestra on frontal apotome, and absence of dorsal mandible tooth. Its remains can indicate presence of Bryozoa colonies (where they live) and consequently stable water level, non-acidified and well-oxygenated conditions.

Harnischia Kieffer (Fig. 6.10)
Yellowish-brown **head** capsule with brown **occipital margin** (Fig. 6.10a). **Antenna** 5-segmented with segments 2 and 3 about equal in length. **Premandible** with 2 large apical teeth and 4 inner teeth decreasing in size to basal tooth. Apical **mandibular** tooth long, 2 blunt inner teeth, divided by short notch, dorsal tooth absent (Fig. 6.10c). **Mentum** with broad, medially notched median tooth, often paler than seven pairs of lateral teeth forming a slightly concave outline. **Ventromental plate** about ¾ the width of mentum and at most weakly striated (Fig. 6.10b).

Remarks
Harnischia subfossils resemble *Paracladopelma* (Fig. 6.16), but in *Paracladopelma* the ventromental plates are longer and coarsely striated and the mentum is rather convex, with lateral teeth not forming a concave curve as in *Harnischia* (Fig. 6.16c). Moreover, in *Paracladopelma* the inner mandibular teeth are pointed and distinctly separated (Fig. 6.16b), and antennal segment 2 is clearly longer than segment 3.

Distribution
Remains were only found in reservoirs in the Štiavnické vrchy Mts. (SK), where they were common but never abundant.

Ecology and palaeoecological interpretation
Larvae of *Harnischia* occur in lakes and slow flowing streams and rivers, preferring sandy bottoms, sometimes with gravel and detritus or mud. Three species are known form Europe (Spies & Sæther 2013), but their separation in larval stage is not possible (Vallenduuk 2017). Larvae of *H. curtilamellata* (Malloch, 1915) live predominantly in the littoral of pools and lakes, but when oxygen content is sufficient, they can also occur in the profundal (at > 20 m depth) (Moller Pillot 2009).

Harnischia remains are rare in stratigraphic records. They can be associated with mesotrophy, non-acidified conditions and relatively high oxygen content.

Lauterborniella Thienemann & Bause (Fig. 6.11)
Head capsule small, yellow; **occipital margin** darker (Fig. 6.11a). **Antenna** 6-segmented, large Lauterborn organs alternate on segments 2 and 3. **Premandible** with 3 distinct apical teeth and 1 blunt inner tooth. **Mandible** pale

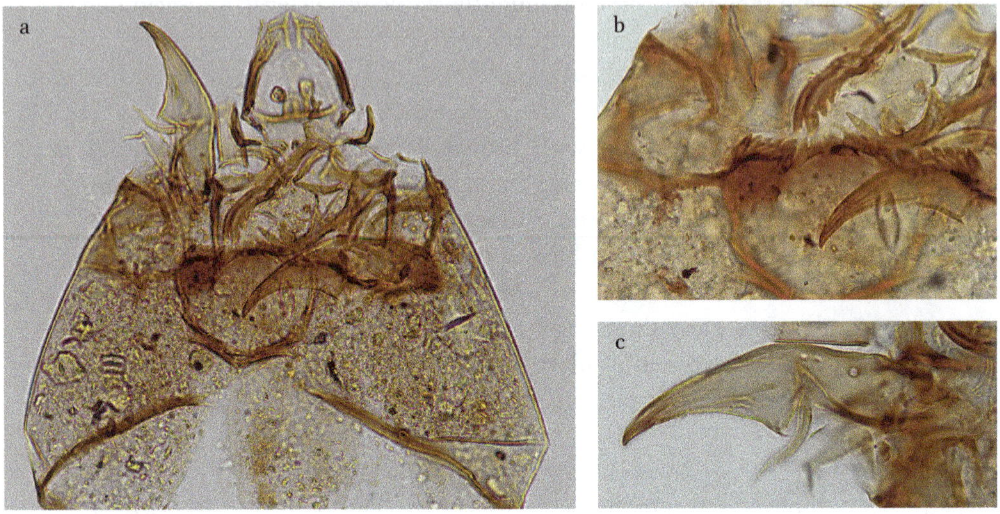

FIGURE 6.10 *Harnischia*. **a** head capsule, **b** mentum, **c** mandible

brown in color; dorsal tooth very strong, apical tooth only slightly larger than 2 inner teeth. **Mentum** weakly pigmented with two rounded median teeth and 6 pairs of lateral teeth; first lateral tooth very short. **Ventromental plates** distinctly triangular in shape (Fig. 6.11b).

Remarks

Lauterborniella may be confused with *Zavreliella* (not shown in this book) and perhaps with *Polypedilum nubeculosum*-type (Figs. 6.19a,b,c). The first and the second lateral teeth of mentum are distinctly separate in *Lauterborniella*, while partially fused in *Zavreliella*. Shape of setae submenti (if preserved in subfossils) is also indicative (branched in *Lauterborniella*, simple in *Zavreliella*). Ventromental plates in *Lauterborniella* are triangular (Fig. 6.11b), while fan-shaped in *Polypedilum nubeculosum*-type (Fig. 6.19b).

Distribution

Subfossil remains were found in one dystrophic Tatra Mts. lake (PL) situated in the forest zone.

Ecology and palaeoecological interpretation

Larvae of the only species known from the Holarctic, *L. agrayloides* (Kieffer, 1911) live in the littoral vegetation of oligotrophic lakes. The species is considered acidophobic (Raddum & Sæther 1981), nevertheless, we found subfossils in a dystrophic Tatra Mts. lake with pH ≤ 5.5 (Bitušík et al. 2006).

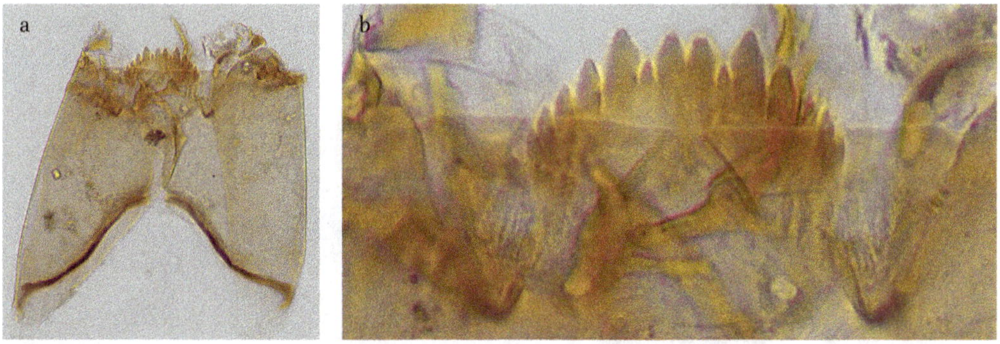

FIGURE 6.11 *Lauterborniella*. **a** head capsule, **b** mentum (note the specific shape of ventromental plates)

Lauterborniella could be associated with changes in lake trophic status and oxygen saturation. However, due to its rarity in subfossil material, it is rarely used for palaeolimnological interpretations.

Microtendipes Kieffer (Fig. 6.12)
Head capsule brownish with dark brown **occipital margin** (Fig. 6.12a). **Antenna** 6-segmented, Lauterborn organs alternate on segments 2 and 3. **Premandible** with 3 (*pedellus*-type) or 5 (*rydalensis*-type) teeth. **Mandible** with pale dorsal tooth, apical tooth and 3 inner teeth darkly pigmented (Fig. 6.12a). Median tooth of **mentum** trifid, pale or as dark as rest of mentum; central tooth either very small, so median tooth appears to consist of two teeth (*pedellus*-type), or about the same size as outer median teeth (*rydalensis*-type); 6 pairs of lateral teeth, first and second laterals fused (Fig. 6.12b). **Ventromental plates** narrow and strongly curved, upper half coarsely striated (Fig. 6.12b).

Remarks
The mentum with 3, usually pale, median teeth (may appear as double with a broad median gap) and mandible with 3 inner teeth are distinctive for *Microtendipes*. Superficially similar genera, such as *Omisus* (Fig. 6.13) and *Paratendipes* (Fig. 6.17) have 4 median mental teeth. *Stictochironomus* is also relatively similar and can be distinguished from *Microtendipes* by the first lateral tooth of mentum that is not shorter than second (Fig. 6.21b).

Morphotypes
Subfossil remains belonged to one morphotype.

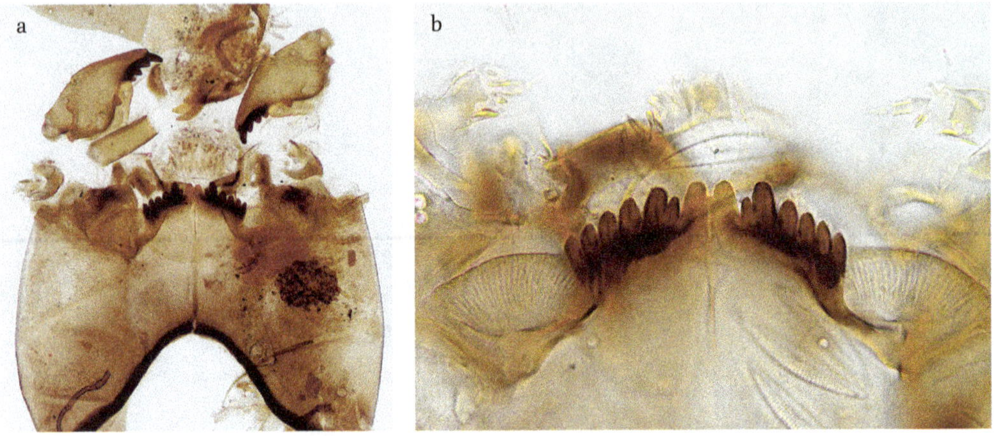

FIGURE 6.12 *Microtendipes pedellus*-type: **a** head capsule, **b** mentum

***Microtendipes pedellus*-type** (Fig. 6.12a). The mentum has a trifid, weakly pigmented median tooth, with the central tooth being very small and barely visible (Fig. 6.12b), premandible with 3 teeth.

Distribution

Subfossil remains were recorded in oligotrophic, non-acidified and warmer subalpine lakes in the Tatra Mts. (SK, PL), Morské oko lake and reservoirs in the Štiavnické vrchy Mts. (SK).

Ecology and palaeoecological interpretation

Larvae of *Microtendipes* are bottom dwellers living in tubes on organic and/or mineral bottoms. While larvae of *M. pedellus* aggregate and *M. rydalensis* (see Vallenduuk 2019 for systematics and nomenclature) live nearly exclusively in flowing waters, larvae of the *chloris* aggregate prefer the littoral zone of lakes (Moller Pillot 2009). Larvae of the genus are considered indicators of oligo/mesotrophic conditions (Brodersen et al. 1998) and are presumably sensitive to oxygen shortage (Rossaro et al. 2007). However, Brodin (1986) listed *Microtendipes* among species characteristic for eutrophic conditions. Even though the information about the relationship of the genus to pH is rather contradictory, most data show the preference of pH values of > 6.0 (Moller Pillot 2009).

The genus is closer to the warm end of the temperature gradient (together with *Chironomus* and *Tanypus*) in the European datasets (Eggermont & Heiri 2011). In the sediment sequence it may signalize intermediate temperatures similarly to northern European lakes where it is frequent during the warmer periods of the Holocene and Lateglacial sediments (Brooks & Birks 2001).

Omisus Townes, 1945 (Fig. 6.13)
Head capsule medium to large, pale with brown **occipital margin** (Fig. 6.13a). **Antenna** 6-segmented. **Premandible** bifid. **Mandible** with short brown apical tooth and 2 inner teeth (3rd inner tooth may be indicated at the base), apical tooth pale (Fig. 6.13c). **Mentum** characteristic: 4 median teeth, central pair narrow and sunken, outer medians broad and tall, somewhat sloping inwardly; 6 pairs of lateral teeth present, 1st lateral shorter than 2nd lateral tooth (Fig. 6.13b). **Ventromental plates** large, broader than mentum, striation obvious only on the basal half (Fig. 6.13b).

Remarks
The shape of mentum resembles *Microtendipes* (Fig. 6.12), with a similar pattern of first and second lateral teeth, but *Omisus* has four median teeth and *Microtendipes* only three. *Paratendipes* (Fig. 6.17) is also similar, but it has all four median teeth subequal in size. In *Stictochironomus*, there are also four median teeth (the middle two reduced), however, the outer median teeth are much longer than laterals, and first and second lateral teeth are not fused.

Morphotypes
The recorded remains belong to one morphotype.

Distribution
Subfossil remains were recorded in small numbers in a dystrophic forest lake in the Tatra Mts. (SK).

Ecology and palaeoecological interpretation
Larvae of the only described Palaearctic species, *Omisus caledonicus* (Edwards, 1932) occupy small dystrophic waterbodies and peat bogs (Epler et al. 2013, own observations).

In the sediment record *Omisus* may indicate paludification of small waterbodies or change from oligotrophy to dystrophy.

Pagastiella Brundin (Fig. 6.14)
Small pale **head capsule** with slightly darker **occipital margin** (Fig. 6.14a). **Antenna** 5-segmented. **Premandible** trifid, first and third longer and pointy, middle tooth shorter and broader. **Mandible** very slender with 2 dorsal and 5 inner teeth; all teeth pale (Fig. 6.14c). **Mentum** arched, weakly pigmented

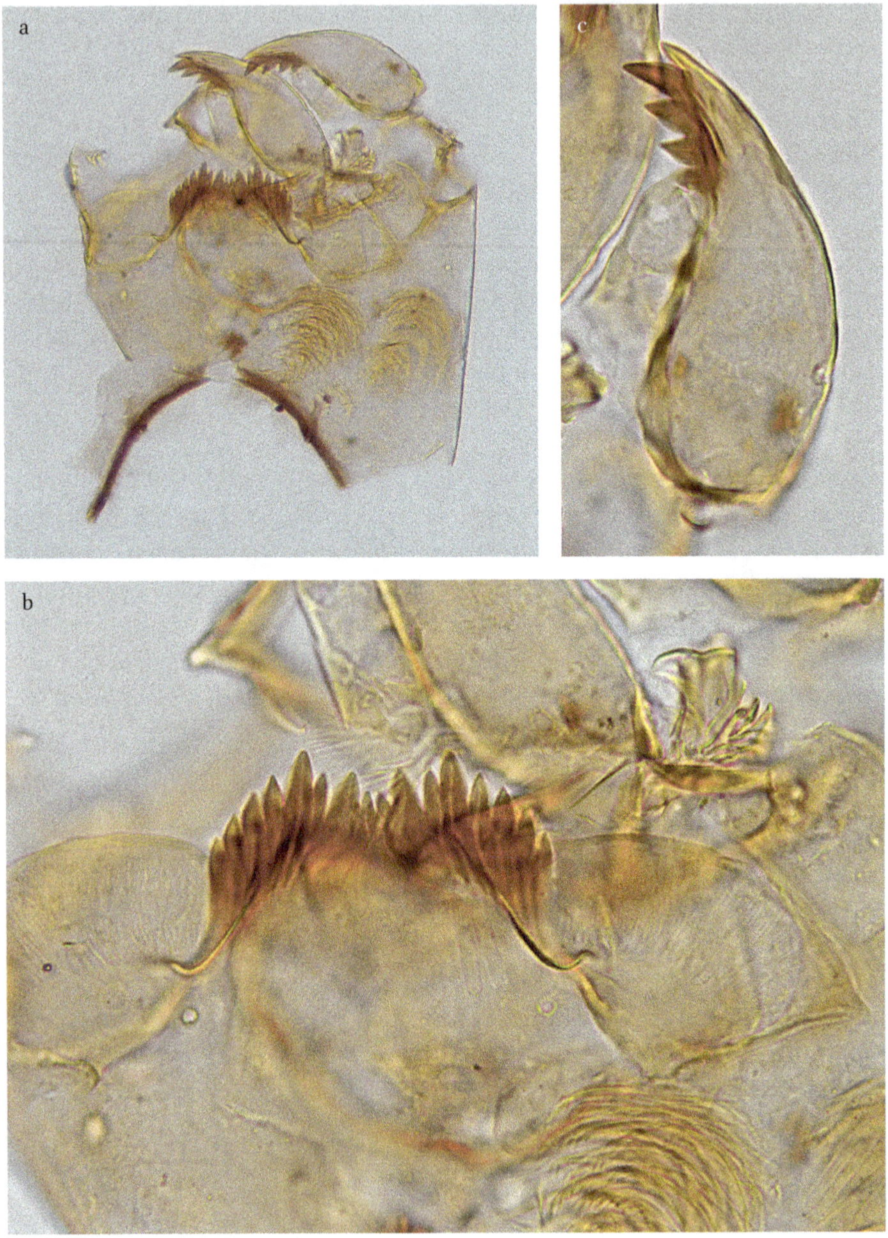

FIGURE 6.13 *Omisus*. **a** head capsule, **b** mentum and ventromental plates, **c** mandible

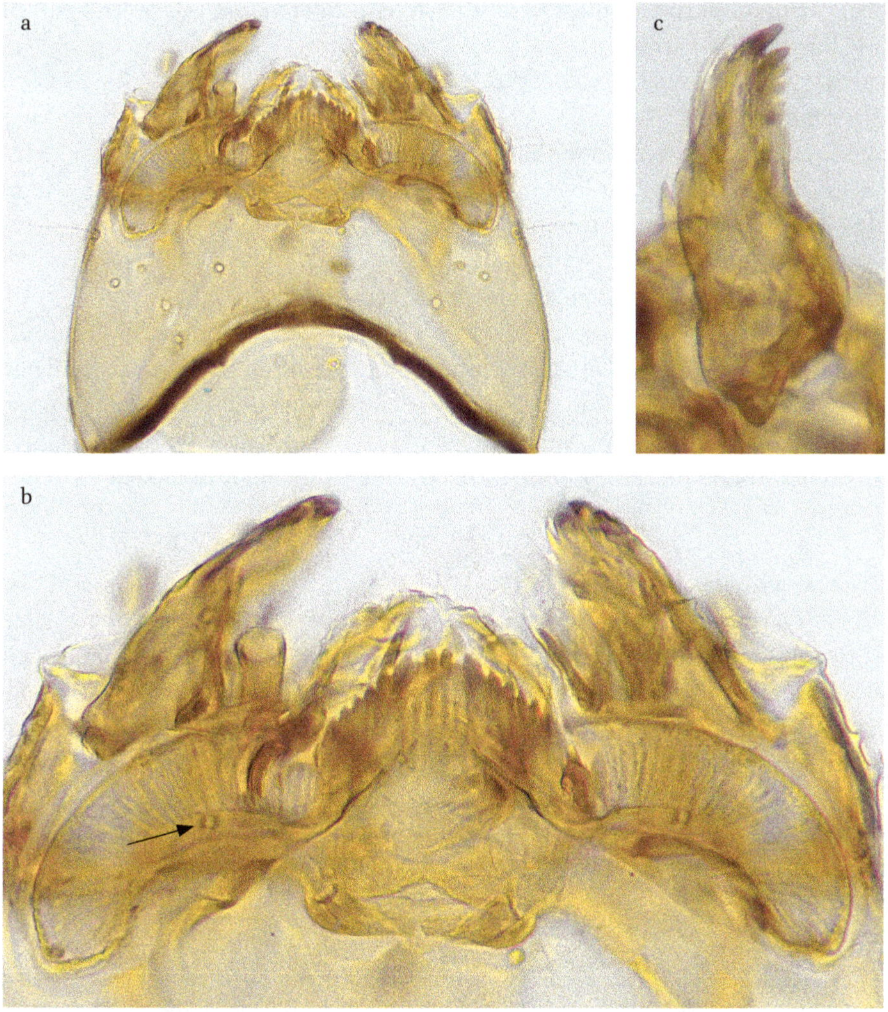

FIGURE 6.14 *Pagastiella*. **a** head capsule, **b** mentum and ventromental plates (arrow indicates position of setae submenti), **c** mandible

with four median teeth and 6 pairs of lateral teeth; outer pair of median teeth very small, first lateral tooth broad and the other laterals decreasing in size (Fig. 6.14b). **Ventromental plates** broad and strongly arched, anterior margin finely crenulated (Fig. 6.14b).

Remarks

Pagastiella differs from *Nilothauma* (not shown in this book) by having unusual position of setae submenti on the ventromental plates (Fig. 6.14b) and very small outer pair of median teeth, while in *Nilothauma* the central pair of median teeth is narrower.

Distribution

Subfossil remains were found rarely and in small numbers in some Tatra Mts. lakes (SK, PL) situated at lower altitudes, mainly in spruce forest zone, and in an alpine lake in the Chornohora Massif (UA).

Ecology and palaeoecological interpretation

Only one species, *P. orophila* (Edwards, 1929), is known from the Palaearctic. Larvae are bottom dwellers of the littoral and sublittoral zone of lakes. They are characteristic inhabitants of oligotrophic and oligo-, mesohumic to polyhumic lakes (Sæther 1979), very often at pH 4–5 (Raddum & Sæther 1981).

Subfossil remains seem to be indicative of humic conditions with low pH (< 5.5).

Parachironomus Lenz (Fig. 6.15)

Head capsule yellow, **occipital margin** ventrally brown, elsewhere pale (Fig. 6.15a). **Antenna** with 5 segments, basal segment robust, longer than flagellum. **Premandible** with 2–4 teeth (Fig. 6.15b). Apical **mandibular** tooth long, 2 inner teeth pointed or flattened, dorsal tooth absents (Fig. 6.15b). Median tooth of **mentum** simple or medially notched, 1.5–2.5× broader than first lateral tooth, and 6–7 pairs of lateral teeth; sixth lateral may be taller than fifth, outermost lateral very small (Fig. 6.15b). **Ventromental plates** about as wide as mentum, anterior margin scalloped to distinctly crenulated (Fig. 6.15c).

Remarks

Pecten epipharyngis (characteristic shape of which is sufficient to distinguish *Parachironomus* larvae from other Chironomini) is usually absent from subfossil heads. Thus, the diagnosis must rely on the shape of mentum and triangular ventromental plates, usually with visibly scalloped anterior margin that is indicative of the genus.

Morphotypes

One morphotype was recognized sensu Brooks et al. (2007).

***Parachironomus varus*-type** (Fig. 6.15a). Median tooth of mentum single and pointed. Mentum slightly convex, sixth lateral tooth higher than fifth, anterior margin of ventromental plates scalloped (Figs. 6.15b,c).

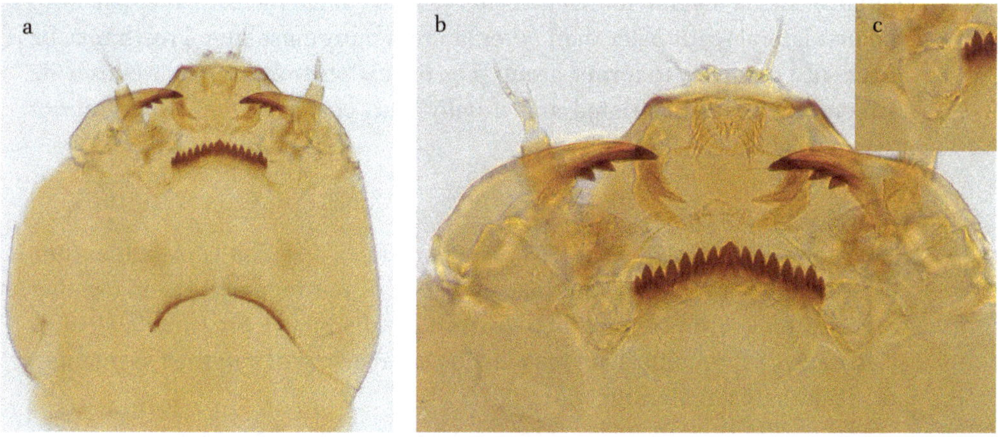

FIGURE 6.15 *Parachironomus varus*-type: **a** head capsule, **b** detail of head capsule showing mentum, ventromental plates, mandibles, premandibles and antennae, **c** detail of ventromental plates (note the scalloped anterior margin)

Distribution

Subfossil remains were rare in surface sediments, only found in a mesotrophic reservoir in the Štiavnické vrchy Mts. (SK).

Ecology and palaeoecological interpretation

Parachironomus larvae occur in various types of stagnant and flowing waters. In spite of general opinion, larvae do not live in soft sediments nor are they true miners of plant tissues. They can be found on macrophytes and firm substrates, including artificial ones where larvae attach their tubes. Some species parasitize on snails or live in bryozoan colonies. In general, the genus is thermophilous and acidophobic (see Moller Pillot 2007 and references therein).

Occurrence of larval subfossils in palaeomaterial may be associated with mesotrophy, and oxygenated littoral with macrophytes and/or hard surfaces, such as stones, pebbles and gravel. Reliability of *Parachironomus* remains for indicating lake trophy and oxygen level seems to be ambiguous (e.g., Langdon et al. 2006).

Paracladopelma Harnisch (Fig. 6.16)

Head capsule brown with dark brown **occipital margin** (Fig. 6.16a). **Antenna** 5-segmented. **Premandible** with 4–7 apical teeth. Apical **mandibular** tooth long, 2 inner teeth pointed and divided by distinct gap, dorsal tooth absents (Fig. 6.16b). **Mentum** with broad to very broad (sometimes dome-shaped) double or single median tooth and usually 7 pairs of lateral teeth; median tooth and first lateral teeth paler than outer laterals, outermost lateral teeth may be larger with tendency to form a group (Fig. 6.16c). **Ventromental plates** as wide as mentum, coarsely striated and usually with coarsely crenulated anterior margin (Fig. 6.16c).

Remarks

Paracladopelma most resembles *Harnischia* (Fig. 6.10). The convex mentum, coarse striation and strong anterior crenulation of ventromental plates, together with the shape of inner mandibular teeth are distinctive for *Paracladopelma* (Figs. 6.16b,c). If antennae are present, the comparison of length of segments 2 and 3 may assist in identification.

Distribution

Subfossil remains were rare, only recorded in reservoirs in the Štiavnické vrchy Mts. (SK).

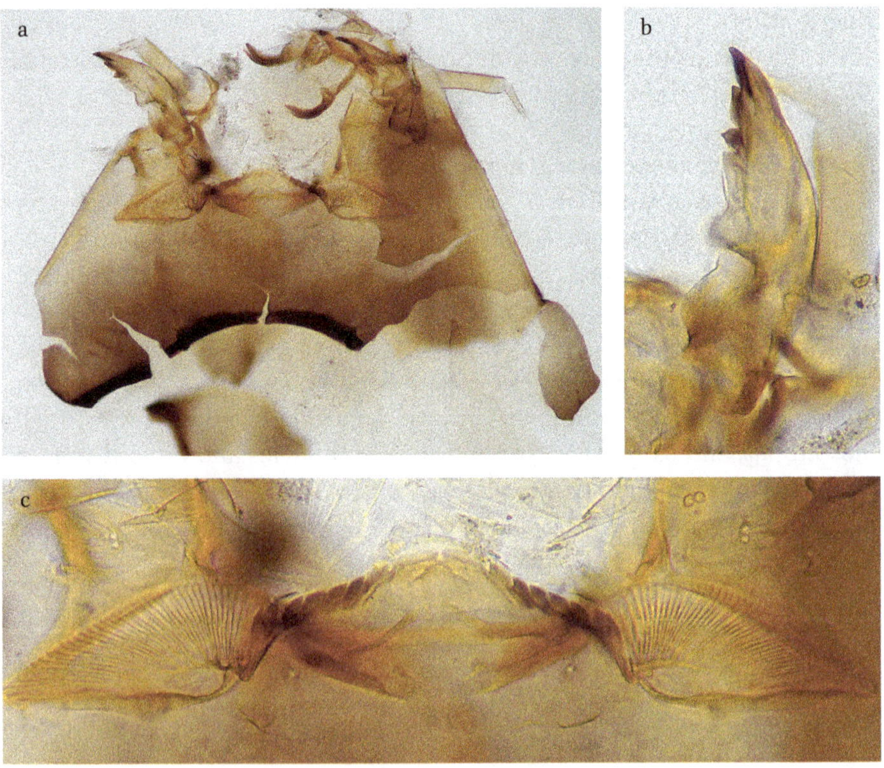

FIGURE 6.16 *Paracladopelma*. **a** head capsule, **b** mandible, **c** mentum

Ecology and palaeoecological interpretation
Paracladopelma is relatively cold-stenothermic (Brodersen & Anderson 2002, Boggero et al. 2006) and requires high concentration of dissolved oxygen. Larvae mostly inhabit sandy substrata in lakes, streams and small rivers and the soft profundal sediments of deep lakes. As obligate predators, they occur in low densities (see Moller Pillot 2007 and references therein).

In the subfossil record, the presence of *Paracladopelma* remains can be associated with well oxygenated littoral and profundal, and oligotrophy to mesotrophy.

Paratendipes Kieffer (Fig. 6.17)
Head capsule yellowish-brown, dorsally with more or less delimited dark areas, **occipital margin** dark brown (Fig. 617a). **Antenna** 6-segmented with alternate Lauterborn organs. **Premandible** with 2–3 teeth. **Mandible** with strong pale apical tooth and 2 inner teeth. **Mentum** with 4, usually pale median teeth, central pair of median teeth equal to or taller than outer pair and with 6 pairs of lateral teeth. First lateral small, sometimes smaller than median teeth and fused with second lateral tooth, which is taller than median teeth; remaining lateral teeth decreasing gradually. Mentum is divided visibly into ventromentum consisting of median teeth and first two pairs of laterals and dorsomentum with 4 pairs of laterals (Fig. 6.17b). **Ventromental plates** widely separated medially, subequal in width to mentum, curved and coarsely striated (Fig. 6.17b).

Remarks
The shape of mentum resembles that of *Microtendipes* (Fig. 6.12) and *Omisus* (Fig. 6.13). *Microtendipes*, however, has only three median teeth (central one hardly visible, fig. 6.12b); *Omisus* has a mentum with four median teeth, but they are all dark and the two central teeth are markedly shorter and narrower than outer median teeth (Fig. 6.13b).

Morphotypes
One morphotype was recorded.
***Paratendipes albimanus*-type** (Fig. 6.17a). Median teeth of mentum shorter than 1st and 2nd lateral teeth (Fig. 6.17b), as opposed to *P. nudisquama*-type (Fig. 6.17c, not present in our material), whose 1st lateral teeth are minute, smaller than median and 2nd lateral teeth.

Distribution
Subfossil remains were found in reservoirs in the Štiavnické vrchy Mts. and Morské oko lake (SK).

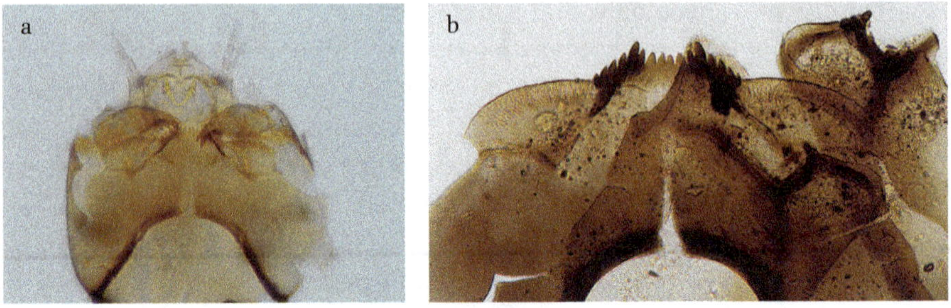

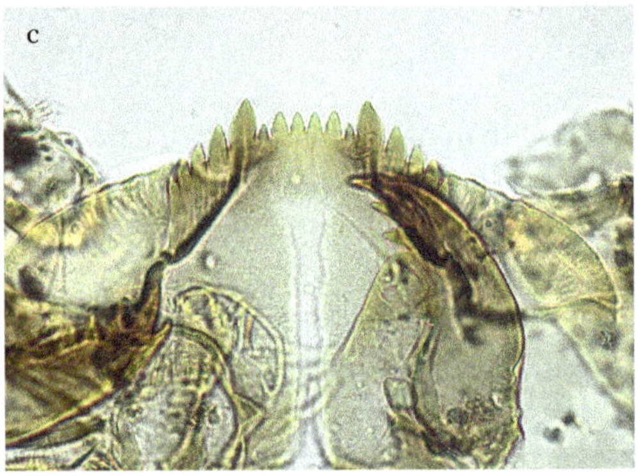

FIGURE 6.17
Paratendipes albimanus-type: **a** head capsule, **b** mentum. **c** *Paratendipes nudisquama*-type mentum
PHOTO HAMERLÍK & SILVA 2021 ©

Ecology and palaeoecological interpretation

Larvae of *Paratendipes* can be found in various types of standing waters (lakes, ponds, small water bodies, peaty moorland pools), streams and rivers. They prefer muddy bottoms but live also on sand and gravel. In lakes larvae occur from littoral to profundal, and oxygen content is probably a limiting factor. *P. nudisquama* (Edwards, 1929) is confined to peaty habitats with pH lower than 5.0, other species seem to be acidophobic (see Moller Pillot 2007 and references therein).

In stratigraphic records, the remains may be indicative of mesotrophic, non-acidified conditions and rather good oxygen supply.

Phaenopsectra Kieffer (Fig. 6.18)

Head capsule medium sized, yellow, **occipital margin** dark brown (Fig. 6.18a). **Antenna** 5-segmented, **premandible** with 2 apical teeth and broad basal tooth. **Mandible** with relatively short dorsal tooth and 3 inner teeth; a deep notch can be present between basal inner tooth and mola; all mandibular teeth dark brown (Fig. 6.18c). **Mentum** darkly pigmented with four median teeth forming ventromentum (set forward of rest of mentum), inner median teeth are shorter than outer teeth; 5–6 pairs of lateral teeth, first lateral tooth much shorter than the second one and outermost pair of lateral teeth reduced (Fig. 6.18b). **Ventromental plates** as long as width of mentum, moderately curved and continuously striated (Fig. 6.18b).

Remarks

Phaenopsectra can be confused with genera that have clearly delineated ventromentum, i.e. *Sergentia* (Fig. 6.20), *Endochironomus/Syndendotendipes* (Fig. 6.8), *Tribelos* (not shown in this book) and *Stictochironomus* (Fig. 6.21). In *Tribelos*, however, the line running from the anterior inner margin of ventromental plates meets the inner median tooth, while in other genera this line meets the outer median tooth. *Sergentia* (Fig. 6.20b) and *Synendotendipes* (Fig. 6.8e) have four inner teeth on the mandible, while *Stictochironomus* usually has only two inner teeth. Ventromental plates in *Endochironomus/Synendotendipes* are extended to a lobe and are longer (Figs. 6.8b,d,f) as opposed to narrower and more curved plates in *Phaenopsectra*.

Morphotypes

One morphotype was identified.

***Phaenopsectra flavipes*-type** (Fig. 6.18a). Mentum strongly arched, with the central pair of median teeth shorter than outer pair, two outermost lateral teeth reduced (Fig. 6.18b). There is a deep notch between basal inner mandibular tooth and the mola (Fig. 6.18c).

Distribution

Subfossil remains were found only in a humic Tatra Mts. lake (SK) situated in the spruce forest zone and in Morské oko lake (SK).

Ecology and palaeoecological interpretation

Phaenopsectra larvae live preferably on plants (stems and leaves) in flowing and stagnant waters. In lakes, larvae live generally in the littoral; however, in oligotrophic to mesotrophic lakes under good oxygen conditions, they survive in the profundal zone, too (Wiederholm & Eriksson 1979). Larvae are evidently tolerant to acidic conditions but their increased abundance in acidified lakes can be a result of cessation of fish predation (Henrikson et al. 1982).

In subfossil records, *Phaenopsectra* remains could indicate extensive littoral with rich macrophyte stands, and/or well oxygenated profundal of oligotrophic/mesotrophic lakes.

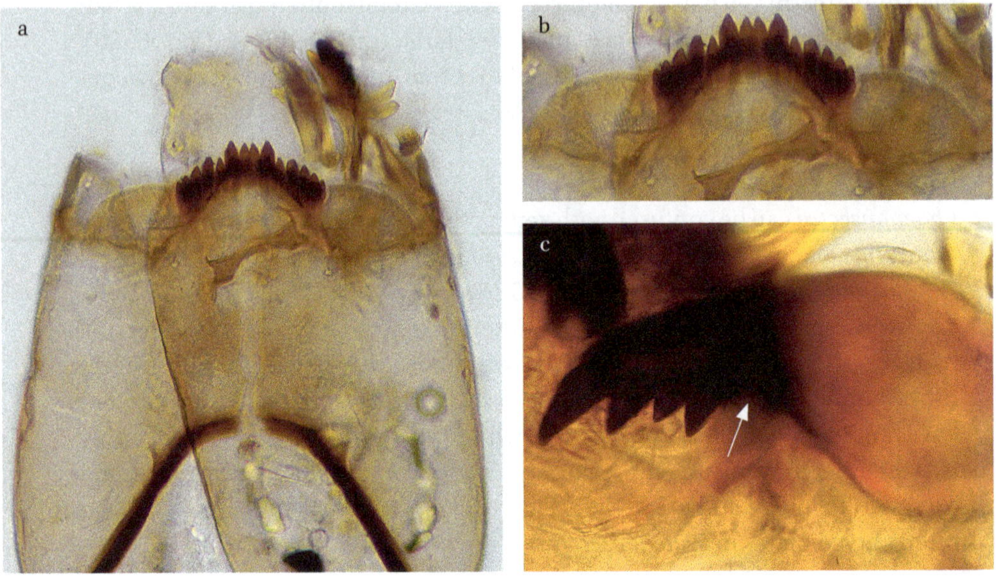

FIGURE 6.18 *Phaenopsectra flavipes*-type: **a** head capsule, **b** mentum, **c** mandible (arrow indicates notch between the basal inner tooth and mola)

Polypedilum Kieffer (Fig. 6.19)
Head capsule small to large, generally yellow, submentum may be darkly pigmented to different extent, usually just above the occipital margin. **Occipital margin** usually dark brown to black, sometimes pale (Figs. 6.19a,d). **Antenna** 5-segmented. **Premandible** with 2 apical teeth and broad basal tooth. **Mandible** generally with 1 dark dorsal tooth and 2 (seldom 3) inner teeth (Figs. 6.19c,f). **Mentum** rather variable in shape, either "*Polypedilum*-like" with 2 tall median teeth, first of 6–7 lateral teeth small, second lateral as tall as median teeth, followed by remainder of short laterals; or all mental teeth subequal, decreasing in size gradually (Figs. 6.19b,e). **Ventromental plates** variable in size and widely separated medially; striae continuous (Figs. 6.19b,e).

Remarks
Larvae of *Polypedilum* are very heterogeneous. Combination of 5-segmented antenna, 2 inner mandibular teeth and dark dorsal tooth together with the typical "*Polypedilum*-like" pattern of mentum (see above) may be sufficient to recognize *Polypedilum*. Species without "*Polypedilum*-like" menta will be distinguished from other genera by 16 subequal mental teeth.

Morphotypes

Two morphotypes were identified following Brooks et al. (2007).

1 Mentum with teeth of various size, median teeth tall, 1st lateral teeth minute and 2nd laterals subequal to median teeth .... *Polypedilum nubeculosum*-type
– Mentum with median and lateral teeth subequal in size .... *Polypedilum sordens*-type

***Polypedilum nubeculosum*-type** (Fig. 6.19a). Mentum "*Polypedilum*-like": median teeth tall, first lateral tooth minute, second lateral tooth as tall as median teeth (Fig. 6.19b); mandible with 2 inner teeth and 1 dorsal tooth (Fig. 6.19c).
***Polypedilum sordens*-type** (Fig. 6.19d). Apex of mentum with teeth subequal in size, median teeth slightly taller and broader than remaining teeth and first lateral teeth only slightly shorter than median and second lateral teeth (Fig. 6.19e). Mandible with 3 inner teeth (in fact, there are only 2 inner teeth, but the robust dorsal tooth seated apically creates the appearance of apical tooth) (Fig. 6.19f).

Distribution

Polypedilum nubeculosum-type was common in the reservoirs of the Štiavnické vrchy Mts. and in Morské oko lake (SK). Small number of subfossils were found in the dystrophic and acidified lakes in the Tatra Mts. (SK, PL) and in one lake in the Chornohora Massif (UA). Subfossil remains of *Polypedilum sordens*-type were recorded scarcely from humic lakes in the Tatra Mts. (SK) and two reservoirs in the Štiavnické vrchy Mts.

Ecology and palaeoecological interpretation

It is impossible to generalize the ecology of *Polypedilum* as it is a large, species-rich and ecologically diverse genus. Moreover, data from literature are, due to incorrect species identification, often rather contradictory. Larval *Polypedilum* live mostly in the upper layers of mud and silt or on gravel, stones and wood; some species (*P. sordens* (van der Wulp, 1874), *P. cultellatum* Goetghebuer, 1931) live on water plants (see Moller Pillot 2007 and references therein). They inhabit flowing and stagnant waters of different types. Some rheophilic species, e.g., *P. albicorne* (Meigen, 1838), *P. convictum* (Walker, 1856), *P. laetum* (Meigen, 1818), *P. pedestre* (Meigen, 1830) and *P. scalaenum* (Schrank, 1803) can also live in the littoral of oxygen-rich lakes. On the other hand, some species occur rather in stagnant water bodies, such as lakes, ponds, ditches, marshes and temporary pools. For example, *P. nubeculosum* (Meigen, 1804), *P. sordens* (van der Wulp,

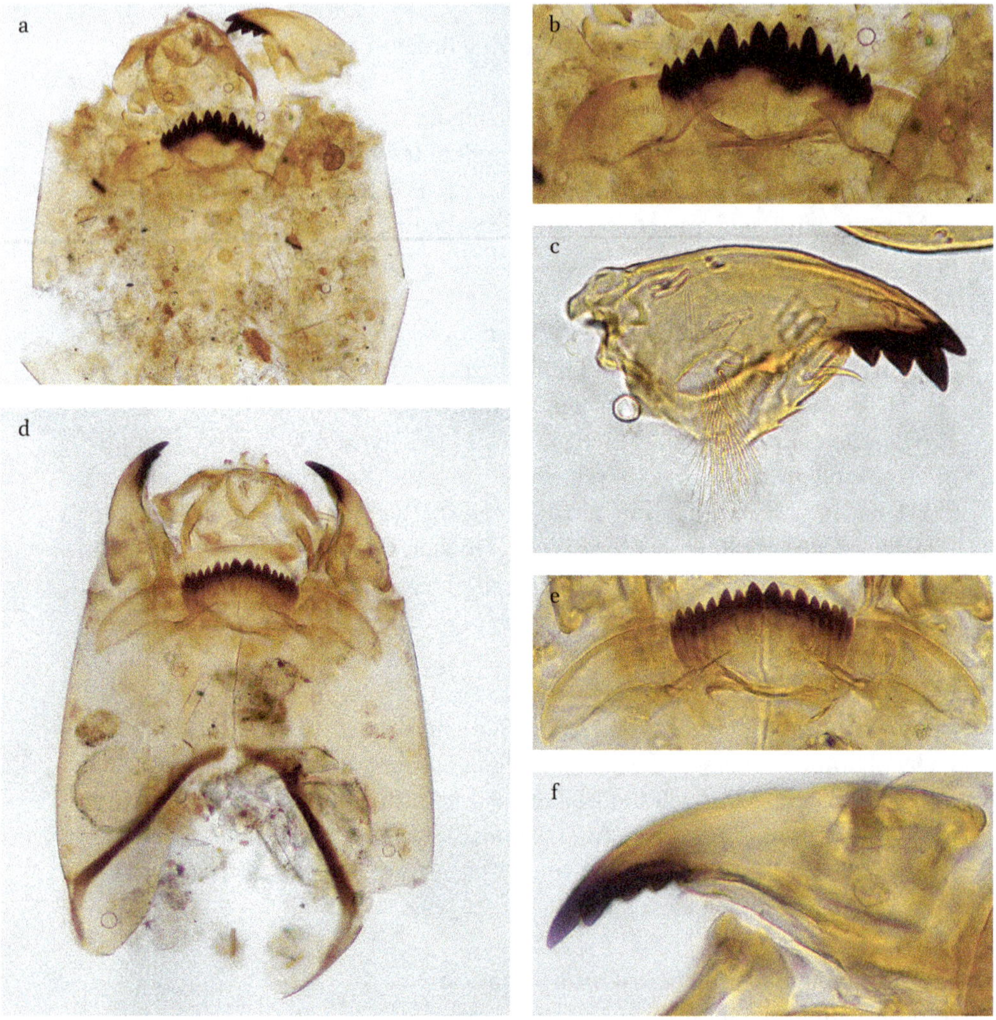

FIGURE 6.19 *Polypedilum. Polypedilum nubeculosum*-type: **a** head capsule, **b** mentum, **c** mandible. *Polypedilum sordens*-type: **d** head capsule, **e** detail of head capsule with mentum and ventromental plates (note their length), **f** mandible

1874), and *P. pullum* (Zetterstedt, 1838) are common inhabitants of lakes. Most *Polypedilum* species prefer waters with high pH, but some were found in water bodies with pH < 5.0 (e.g., *P. pullum* and *P. uncinatum*).

The correct interpretation of *Polypedilum* remains in stratigraphic records should be based on reliable identification. While *P. nubeculosum*-type is indicative of meso- to eutrophic conditions or increased lake trophy (e.g., Brodin

1986, Samartin et al. 2012, Chamutiová et al. 2018), *P. sordens*-type requires oligotrophy and stable oxygen regime, moreover, it is associated with macrophytes (e.g., Brodersen et al. 2001b and references therein).

Sergentia **Kieffer** (Fig. 6.20)
Head capsule large, yellow (Fig. 6.20a). **Antenna** 5-segmented, **premandible** with 2–4 teeth. All **mandibular** teeth dull brown. One relatively short dorsal tooth and 4 inner teeth present (Fig. 6.20b). **Mentum** with 4 large, separated, equally broad and slightly elevated median teeth, central pair slightly shorter than outer one, and with 6 pairs of lateral teeth (Fig. 6.20b). **Ventromental plates** as wide as mentum, moderately curved and continuously striated, in *S. coracina*-type, however, with obvious paired striations along anterior margin (Fig. 6.20c).

Remarks
Sergentia is closely related to *Phaenopsectra* (Fig. 6.18) and the shape of mentum, which is divided into three sections also resembles some other genera (see remarks to *Phaenopsectra*). *Sergentia* may be recognized by the 4 inner mandibular teeth, 4 subequal median mental teeth, not reduced outermost lateral teeth (Fig. 6.20b), and paired striations conspicuous along anterior margin in *S. coracina*-type (Fig. 6.20c).

Morphotypes
One morphotype was recognised following Brooks et al. (2007).
***Sergentia coracina*-type** (Fig. 6.20a). Mandible with 4 dark inner teeth, premandibles with 2 apical teeth, mentum with 4 subequal median teeth, outermost lateral teeth not reduced (Fig. 6.20b), and distinct paired striations along anterior margin of ventromental plate (Fig. 6.20c).

Distribution
Head capsules were found in the sediments of a reservoir in the Štiavnické vrchy Mts., in Morské oko lake and in one shallow alpine Tatra Mts. lake (SK).

Ecology and palaeoecological interpretation
Larvae of the most common European species *Sergentia coracina* (Zetterstedt, 1850) occur on muddy bottom in deeper parts (sublittoral, profundal) of oligotrophic and mesotrophic lakes (Wiederholm 1980). Increased abundance of *Sergentia* may indicate acidification (Brodin & Gransberg 1993, Raddum & Sæther 1981).

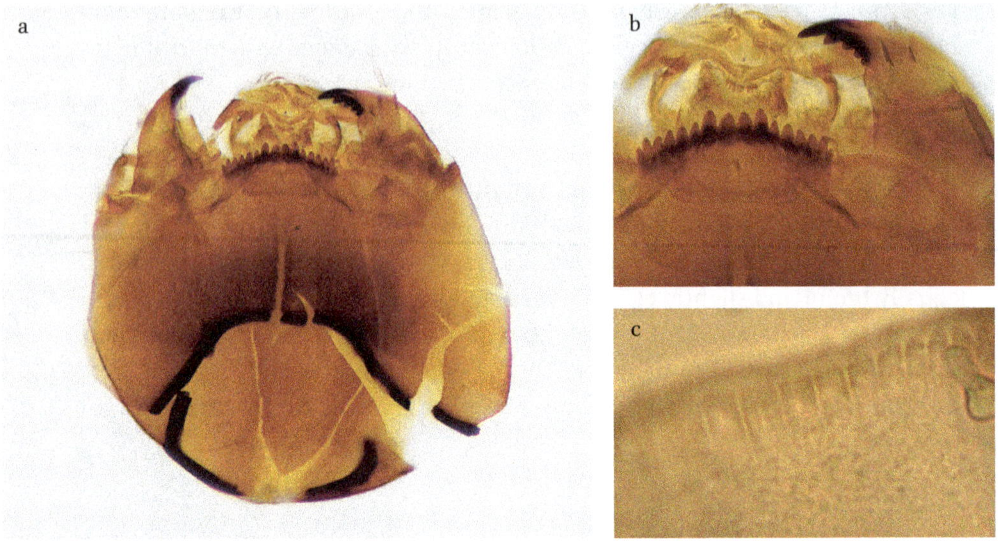

FIGURE 6.20 *Sergentia coracina* – type: **a** head capsule, **b** mentum and mandible, **c** paired striations on the ventromental plate

In palaeolimnological studies, the succession through *Heterotrissocladius*/*Micropsectra*, *Stictochironomus*/*Sergentia* and *Chironomus*/*Procladius* stages is linked to the decrease in hypolimnetic oxygen (see Brodersen et al. 2001b and references therein). Decreasing abundance of *Sergentia* signalizes increased nutrient load (Brodersen & Quinlan 2006).

Stictochironomus Kieffer (Figure 92) (Fig. 6.21)
Head capsule medium to large, yellow, posteriorly may be darkly pigmented, and dorsally with distinct brown areas. **Occipital margin** mostly dark brown (Fig. 6.21a). **Antenna** 6-segmented, Lauterborn organs alternate on segment 2 and 3. **Premandible** with 2 apical teeth and broad basal tooth. **Mandible** basally broad and very slender distally, with 1 dorsal tooth and 2 inner teeth, all mandibular teeth dark. **Mentum** with 4 median teeth delineated from lateral regions; central pair of median teeth lower and slimmer than outer pair; 6 pairs of lateral teeth; all teeth dark (Fig. 6.21b). **Ventromental plates** slightly wider than mentum, weakly curved, striations often indistinct, broad grey band along anterior margin (Figs. 6.21c,d).

Remarks
The shape of mentum with delineated ventromentum resembles some other genera (see comments to *Phaenopsectra*). However, the central pair of median

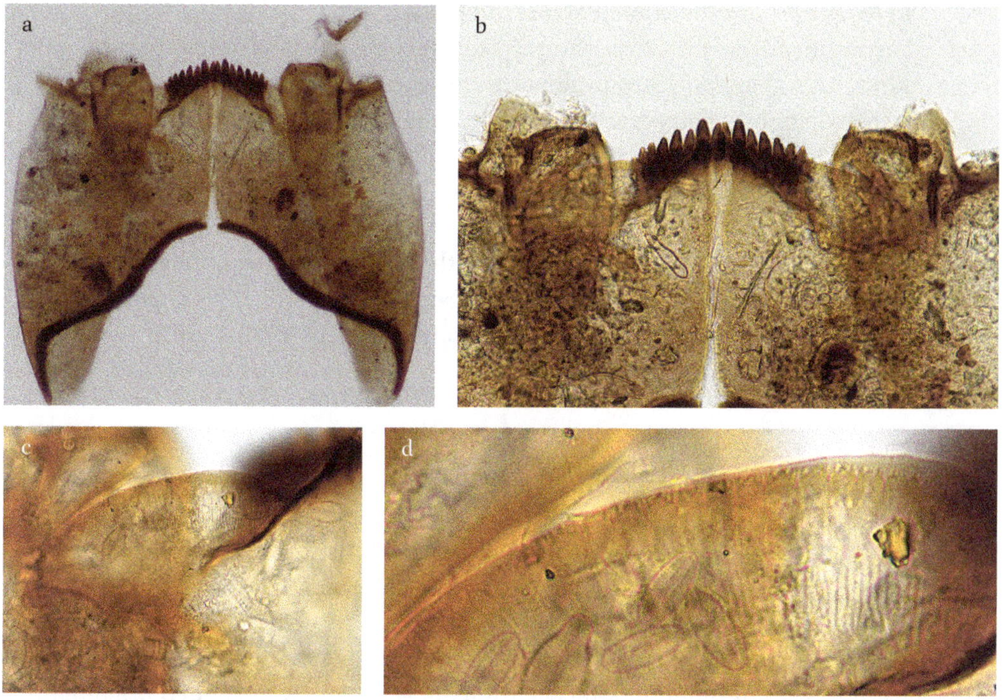

FIGURE 6.21 *Stictochironomus*. **a** head capsule, **b** mentum, **c** ventromental plate, **d** broad grey band along anterior margin of ventromental plate

teeth is thinner in *Stictochironomus* (Fig. 6.21b). Moreover, the 6-segmented antenna with alternate Lauterborn organs (normally absent in subfossils), 2 inner mandibular teeth and the grey band on the anterior margin on ventromental plates are also distinctive (Fig. 6.21d).

Morphotypes

One morphotype was distinguished following Brooks et al. (2007).

***Stictochironomus* type B** (Fig. 6.21a). Even though the morphotype looks slightly different from type B shown in Brooks et al. (2007), we assume that it is the same type. Unfortunately, in our specimen the middle part of the mentum is broken and the central median teeth are superimposed and party covering each other, giving the look of slightly narrower and lower central teeth compared to the original type B (Fig. 6.21b).

Distribution

Few subfossil remains of the same morphotype were found in a large forest lake in the Tatra Mts. (PL).

Ecology and palaeoecological interpretation

Larvae of *Stictochironomus* are typical bottom dwellers inhabiting mostly sandy and silty substrata in flowing and stagnant waters. Some species are rather rheophilic, however *S. crassiforceps* (Kieffer, 1922) and *S. rosenschoeldi* (Zetterstedt, 1838) are confined to lentic systems. Larvae of all species require oxygen-rich environments and Johnson (1989) considered the whole genus indicative of oligo-mesotrophic conditions in Swedish lakes. Especially *S. rosenschoeldi*, as a typical inhabitant of the profundal zone of oligotrophic to mesotrophic lakes, is an important indicator of hypolimnetic oxygen content both in recent limnological and palaeolimnological studies (also see comments to *Sergentia*).

6.2 Tribe Tanytarsini

Head capsule pale to brown. Occipital margin darkened, dark brown to black. Shape and degree of development of occipital plate, if present, is important for identification of subfossils. Antenna always 5-segmented arising from a distinct elongate antennal base – pedestals, which are distally often projecting into spur of different size, occasionally accessory multispined or palmate process is present. Lauterborn organs well-developed, usually mounted on segment 2 and often situated on short to long pedicels. Premandible with 2–5 teeth. Mandible with 2–3 inner teeth, one dorsal tooth (occasionally 2–3 dorsal teeth); 1–2 surficial teeth may be present on dorsal surface. Mentum with single median tooth (may be laterally notched) and mostly with 5 lateral teeth. Ventromental plates usually wide and slender, almost touching each other medially, in some genera narrow, separated medially by at least the width of the three median teeth.

KEY TO GENERA OF SUBFOSSIL TANYTARSINI

1 Antennal pedestal with large multi-branched spur .... *Stempellina*
– If spur is present on antennal pedestal, it is simple and short, at most ¼ the lengths of pedestal 2

2 Second lateral tooth of mentum distinctly shorter than 1st and 3rd laterals .... *Cladotanytarsus*
– Second lateral tooth of mentum subequal to adjacent lateral teeth 3

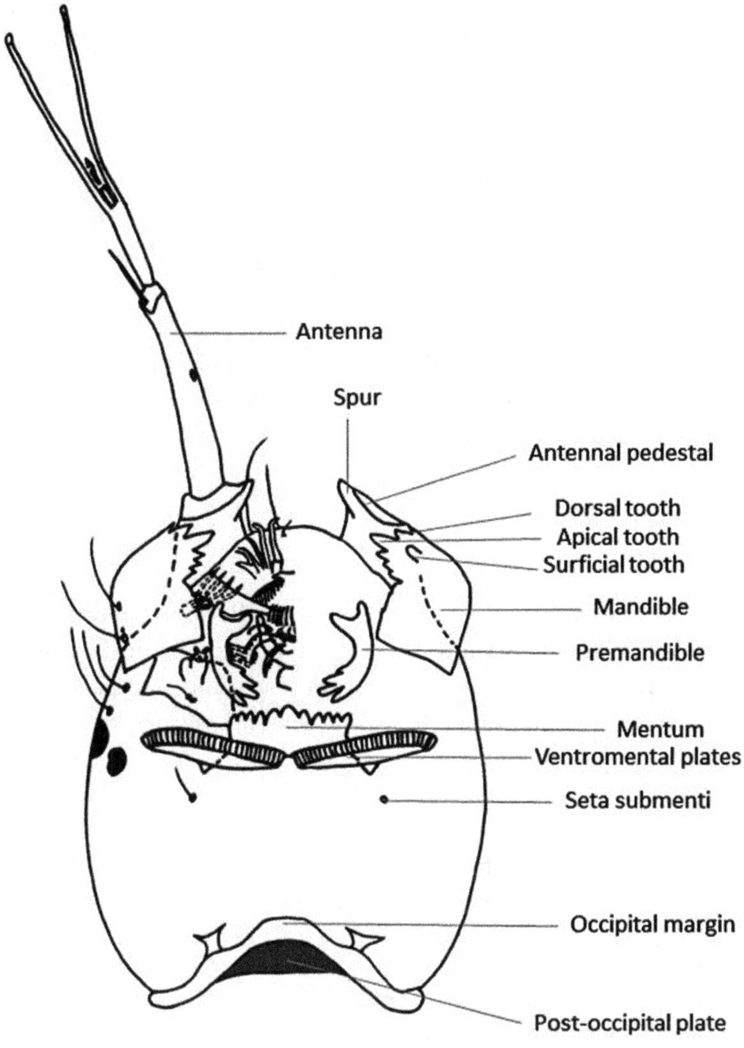

FIGURE 6.22 Tanytarsini larval head capsule showing subfossil diagnostic characters, comparing recent head (left side) and a subfossil remain (right side). Redrawn and modified after Oliver et al. (1978)

3 Antennal pedestal with spur .... *Micropsectra/Tanytarsus* in part
– Antennal pedestal without spur 4

4 Antennal pedestal short, about as long as wide, mentum equally colored .... *Paratanytarsus*/Tanytarsini indet.

– Antennal pedestal longer, about 2× as long as wide, central part of mentum usually lighter than remainder .... *Tanytarsus* in part

Cladotanytarsus Kieffer (Fig. 6.23)
Head capsule yellow, ventrally brownish; **occipital margin** ventrally dark brown, elsewhere slightly darker than head capsule (Fig. 6.23a). **Antenna** 5-segmented, placed on short pedestal without apical tooth or spur; segment 2 short, wedge shaped. Lauterborn organs large, placed on short pedicels (Fig. 6.23b). **Premadible** with 4–5 teeth. **Mandible** with pale dorsal tooth, apical tooth and 2–3 inner teeth pale brown to brown (Fig. 6.23b). Median **mental** tooth broad, rounded or notched laterally; 5 pairs of lateral teeth gradually decreasing in size, or second laterals much smaller than adjacent, or first laterals small and closely pressed to median tooth so mentum appears to consist only of 4 pairs of separate lateral teeth (Fig. 6.23c). **Ventromental plates** long and narrow, close together medially (Fig. 6.23c).

Remarks
The most characteristic features of the genus are the short, wedge shaped 2nd antennal segment and large Lauterborn organs on short pedicels. Provided that antennae are missing, the shape of mentum (especially in case of *C. mancus*-type with minute 2nd lateral teeth), combined with narrow and long ventromental plates, may be distinctive.

Morphotypes
One morphotype was identified.
***Cladotanytarsus mancus*-type** (Fig. 6.23a). The mentum with distinctly shorter second lateral tooth is distinctive not only within the genus but for Tanytarsini in general (Fig. 6.23c).

Distribution
Cladotanytarsus remains were frequent in sediments of the reservoirs in the Štiavnické vrchy Mts. (SK); they were also found in Morské oko lake (SK) and in about half of the lakes in the Svydovets Massif (UA).

Ecology and palaeoecological interpretation
Larvae are ecologically versatile, living in a variety of lentic and lotic habitats, including brackish water and hot springs. They inhabit predominantly soft sediments with sand in the littoral zone and probably avoid macrophyte vegetation. Larvae can be found in lakes ranging from oligotrophy to eutrophy, however, they are intolerant to anoxic conditions (van de Bund et al. 1995, van

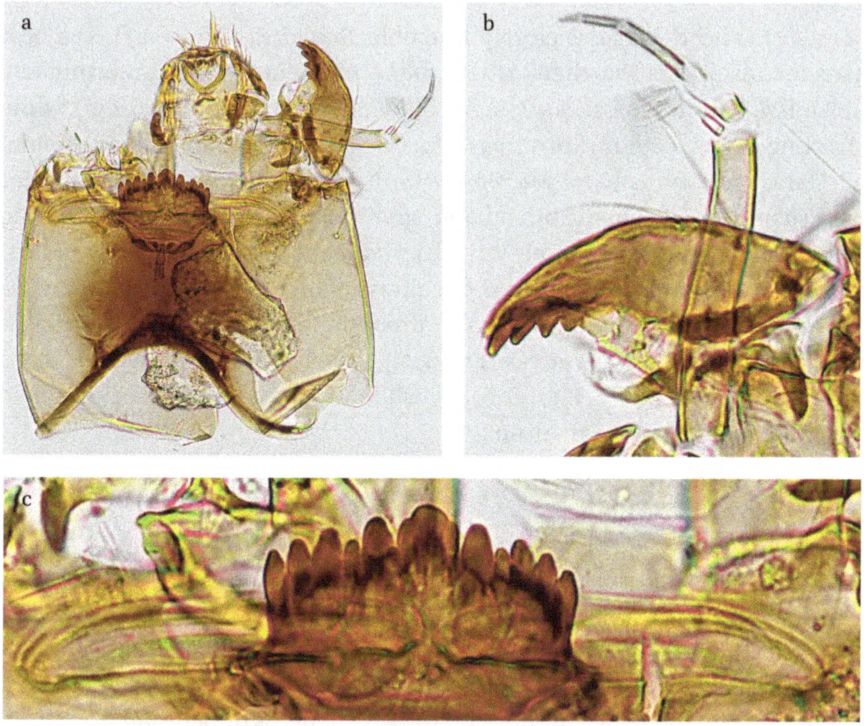

FIGURE 6.23 *Cladotanytarsus mancus*-type: **a** head capsule, **b** mandible and antenna, **c** mentum (note the reduced 2nd lateral tooth)

den Berg et al. 1997). The genus is thermophilous in general, but larvae have been found in lakes in the Svydovets Massif (UA) at altitude ca. 1500 m.

Cladotanytarsus remains may be abundant in stratigraphic records and indicative of relatively warm and well-oxygenated water (Gandouin & Franquet 2002, Laroque & Hall 2003) in mesotrophic, or even eutrophic conditions (Langdon et al. 2006, Samartin et al. 2012).

Micropsectra Kieffer (Fig. 6.24)

Head capsule yellow to brown. **Occipital margin** usually with well-developed **occipital plate** (Figs. 6.24a,d,g,i,j,k). **Antenna** placed on elongate pedestal, often with apical spur. Lauterborn organs on long pedicels that extend beyond flagellum. **Premandible** bifid (Fig. 6.24b). **Mandible** with apical tooth, dorsal tooth and 3 inner teeth, all teeth brown (Fig. 6.24c). **Mentum** with the median tooth rounded, laterally notched to crenate, usually slightly paler than 5 lateral teeth (Figs. 6.24e,h,l). **Ventromental plates** close together medially, a little wider than mentum, with fine striation (Figs. 6.24e,h,l).

Remarks

Micropsectra head capsules closely resemble *Tanytarsus* (Fig. 6.27). The best character for separating them is the shape of premandible (unfortunately often missing from subfossils), which is bifid in *Micropsectra* (Fig. 6.24b), while consisting of 3–5 teeth in *Tanytarsus* (Fig. 6.27g). A projection on the antennal pedestal is characteristic for *Micropsectra*. However, some *Tanytarsus* species may also have a spur. On the other hand, spur is not always visible or is missing in some *Micropsectra* (very rarely though). The shape of spur can be also indicative, since it is sharp in *Micropsectra*, but always of different shape in *Tanytarsus* (e.g., long, broad and rounded apically; Brooks et al., 2007). *Micropsectra* can be confused with *Paratanytarsus* (Fig. 6.25) as well, but *Paratanytarsus* has a uniformly dark mentum (Figs. 6.25c,f), while in *Micropsectra* there is a pale area in the middle part of mentum.

Morphotypes

Six morphotypes were identified in the subfossil material following Brooks et al. (2007).

1 Spur on antennal pedestal long, about 1/3 the lengths of pedestal .... *Micropsectra pallidula*-type
– Spur on antennal pedestal short, up to ¼ the lengths of pedestal 2

2 Head capsule usually dark, antennal pedestal approximately as long as broad, spur very short, post-occipital plate missing or reduced .... *Micropsectra radialis*-type
– Head capsule usually paler, antennal pedestal longer than broad, spur longer, post-occipital plate strongly developed 3

3 2nd lateral mental tooth taller than 1st and 3rd laterals .... *Micropsectra junci*-type
– 2nd and 1st lateral mental teeth subequal in size 4

4 Post-occipital plate convex, projecting posteriorly .... *Micropsectra* type A
– Post-occipital plate concave at most with straight margin 5

5 Post-occipital arch broad, spur straight .... *Micropsectra insignilobus*-type
– Post-occipital arch narrow, spur points outwards .... *Micropsectra contracta*-type

***Micropsectra contracta*-type.** Head capsule pale brown. Arch of occipital margin relatively narrow. Spurs on antennal pedestals distinctive and noticeably point outwards (Fig. 6.24a).
***Micropsectra insignilobus*-type.** Head capsule pale brown. Long antennal pedestals with short, pointed spurs (Fig. 6.24d). Post-occipital plate well-developed, arch of occipital margin relatively broad (Fig. 6.24f).
***Micropsectra junci*-type.** Head capsule yellow (Fig. 6.24g). Similar to *M. insignilobus*-type, but first pair of lateral teeth of mentum clearly shorter than the second laterals (Fig. 6.24h).
***Micropsectra pallidula*-type.** Head capsule dark brown (obviously darker than the one in Brooks et al. 2007). Antennal pedestal about twice as long as broad, spur long, about half the length of pedestal, blunt apically. Post-occipital plate present (Fig. 6.24i).
***Micropsectra radialis*-type.** Head capsule brown. Antennal pedestal about as long as broad, bearing a short rounded spur. Post-occipital plate missing or strongly reduced (Fig. 6.24j).
***Micropsectra* type A.** Head capsule yellow (Fig. 6.24k). Similar to *M. insignilobus*-type, but post-occipital plate is enlarged and convex (Fig. 6.24m). Based on samples where both pupal exuviae and living larvae were recorded (lake Vinderel, Romania, and lakes of the Azores Islands) this morphotype corresponds to *Micropsectra lindrothi* Goetghebuer, 1931 (unpublished data).

Distribution

Subfossil *Micropsectra* remains have been found in many of the studied lakes, especially in the Tatra Mts. (SK, PL), where the highest morphotype diversity was recorded. *Micropsectra radialis*-type was frequent and abundant in the coldest Tatra Mts. lakes. *Micropsectra* type A was the second most abundant morphotype in the Tatra Mts. *Micropsectra* was absent from Morské oko lake (SK), and it was relatively abundant in the reservoirs in the Štiavnické vrchy Mts. (SK).

Ecology and palaeoecological interpretation

Micropsectra is a species-rich genus with larvae generally restricted to cold and well-oxygenated conditions in stagnant and flowing waters. They are typical bottom dwellers, inhabiting muddy substrata, but larvae of some species were found in the moss at strong current (*M. klinki* Stur & Ekrem, 2006) or in hygropetric situations (*M. bavarica* Stur & Ekrem, 2006). Larvae live mainly in slack sections of streams and rivers and in lakes, including the profundal zone, but they have been recorded from a wide range of habitats: springs, wetland pools,

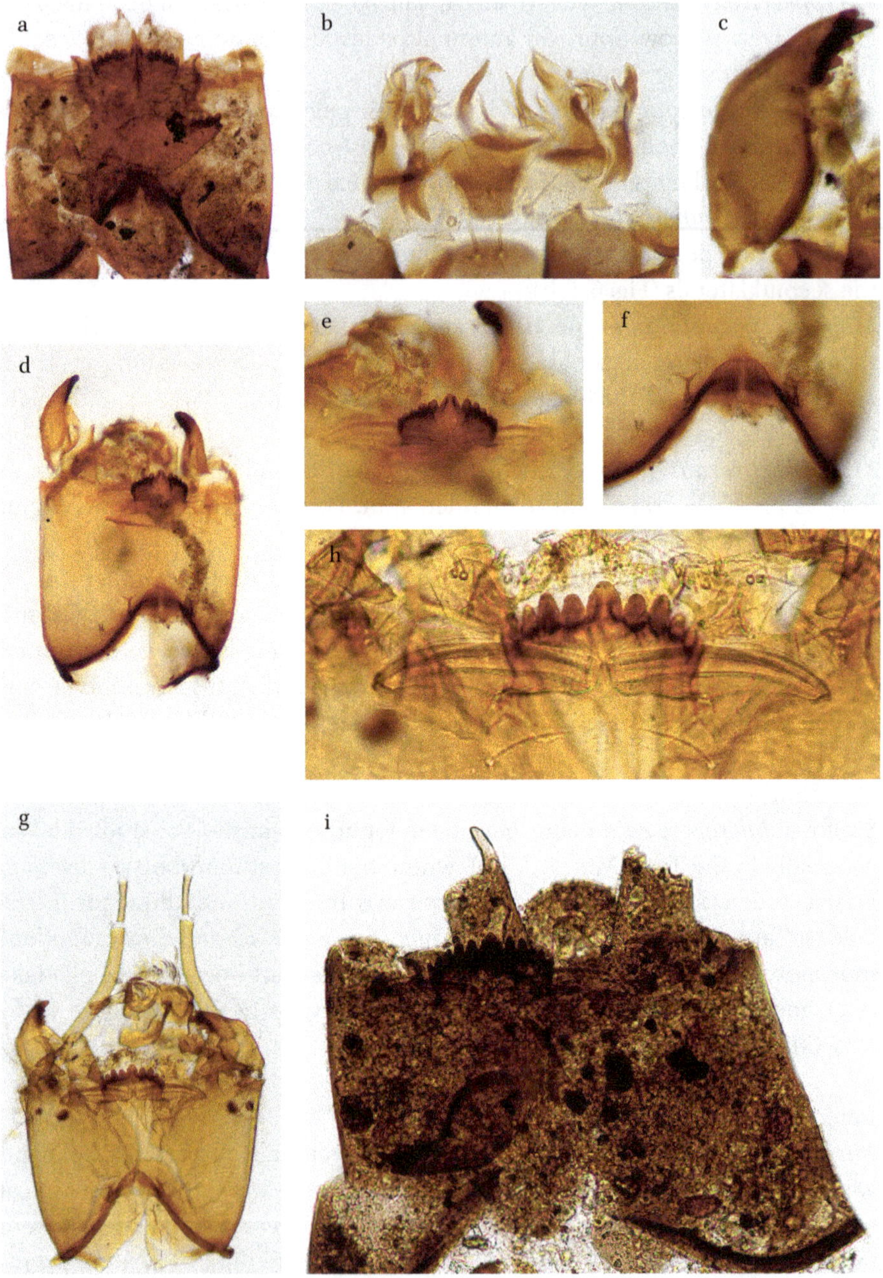

FIGURE 6.24 *Micropsectra*. *Micropsectra contracta*-type: **a** head capsule, **b** detail of head capsule showing bifid premandibles (characteristic for *Micropsectra*), **c** mandible. *Micropsectra insignilobus*-type: **d** head capsule, **e** mentum, **f** detail of occipital margin and post-occipital plate. *Micropsectra junci*-type: **g** head capsule, **h** mentum. *Micropsectra pallidula*-type: **i** head capsule

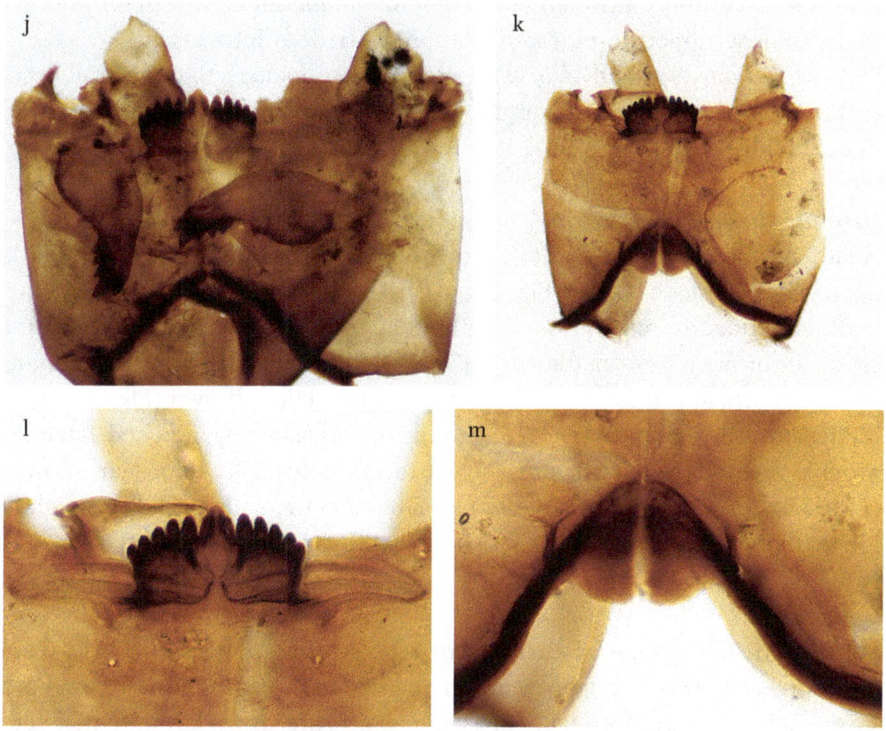

FIGURE 6.24 (continuation) *Micropsectra radialis*-type: j head capsule. *Micropsectra* type A: **k** head capsule, **l** mentum, **m** convex post-occipital plate

ditches, temporary waters (Stur & Ekrem 2006, Epler et al. 2013, own observations). Some species can be dominant in Arctic and high-altitude lakes (Alm & Willassen 1993).

In lakes, *Micropsectra* will indicate oligotrophy (Wiederholm 1980) and together with *Heterotrissocladius* and *Paracladopelma* the genus belongs to genera indicative of cold and unproductive conditions (Brodersen & Anderson 2002). The extreme cold-stenothermal *Micropsectra radialis*-type is a particularly reliable palaeo-indicator of changes in air- and water temperature and lake depth (e.g., Laroque-Tobler et al 2010 and references therein). Together with *Pseudodiamesa nivosa*, *M. radialis*-type occupy the coldest part of the temperature gradient of both the Holarctic training sets (Eggermont & Heiri 2012) and the Tatra training-set (unpublished data). The extirpation of the above mentioned morphotypes from the sediment sequence marked the disappearance of the glacier from the catchment at the Glacial-Holocene boundary (Bitušík et al. 2017). Currently, the combination of *M. radialis* and *P. nivalis* only occurs in the coldest (uppermost located or strongly shaded) Tatra lakes.

As *M. radialis*-type and *M. insignilobus*-type are acidophobic (e.g., Kubovčík & Bitušík 2006), they can be used for the reconstruction of lake acidification (e.g., Stuchlík et al. 2017).

Paratanytarsus Thienemann & Bause (Fig. 6.25)
Head capsules yellow with dark brown **occipital margin. Occipital arch** is deeply incised, plate absent (Figs. 6.25a,d). **Antenna** placed on relatively short pedestal (about as long as broad) without any projections. Lauterborn organs sessile or on very short pedicels (Fig. 6.25a). **Premandible** bifid. Dorsal **mandibular** tooth, apical tooth and 2–3 inner teeth all dark brown (Figs. 6.25b,e). Median tooth of **mentum** usually rounded, occasionally notched laterally, 5 pairs of lateral teeth are present, all teeth uniform in color, dark brown (Figs. 6.25c,f). **Ventromental plates** close together medially, each plate slightly longer than width of mentum (Figs. 6.25c,f).

Remarks
Subfossil head capsules of *Paratanytarsus* resemble *Micropsectra* (Fig. 6.24) and *Tanytarsus* (Fig. 6.27), but *Paratanytarsus* can be distinguished from both by the combination of a deeply incised occipital arc, and short antennal pedestal without a spur (Figs. 6.25a,d). Another useful diagnostic feature is the uniformly colored mentum with all teeth dark (Figs. 6.25c,f), compared to the paler median part of mentum in *Micropsectra* and *Tanytarsus*.

Morphotypes
Two morphotypes were present following Brooks et al. (2007). Their separation requires the presence of mandibles.
***Paratanytarsus austriacus*-type** (Fig. 6.25a). Mandible with 3 inner teeth (Fig. 6.25b).
***Paratanytarsus penicillatus*-type** (Fig. 6.25d). Mandible with 2 inner teeth (Fig. 6.25e).

Distribution
Paratanytarsus remains were common and were recorded in the Tatra Mts. (SK, PL), the reservoirs of the Štiavnické vrchy Mts. (SK), and in the Chornohora and Svydovets Massifs (UA). While *Paratanytarsus austriacus*-type was characteristic for alpine lakes, *Paratanytarsus pencillatus*-type occurred in lakes of lower altitudes, e.g., in the Ukrainian Carpathians, or in dystrophic Tatra Mts. lakes situated in the forest zone.

Ecology and palaeoecological interpretation

Generally, *Paratanytarsus* is a eurytopic genus, with larvae living in a wide range of water types: streams, rivers, lakes, reservoirs, wetland pools and brackish ponds. Larvae of the parthenogenetic species *P. grimmii* (Schneider, 1885) live even in water-supply systems (Epler et al. 2013, Armitage et al 1995, Hamerlík & Brodersen 2010, own observations). In lakes, larvae occur in the littoral and sublittoral; sometimes they prefer macrophyte mats (Brodersen et al. 2001b, Tarrats et al. 2018).

Larval remains are frequent in stratigraphic records. *Paratanytarsus austriacus*-type is associated with low temperature and oligotrophic conditions (Lencioni

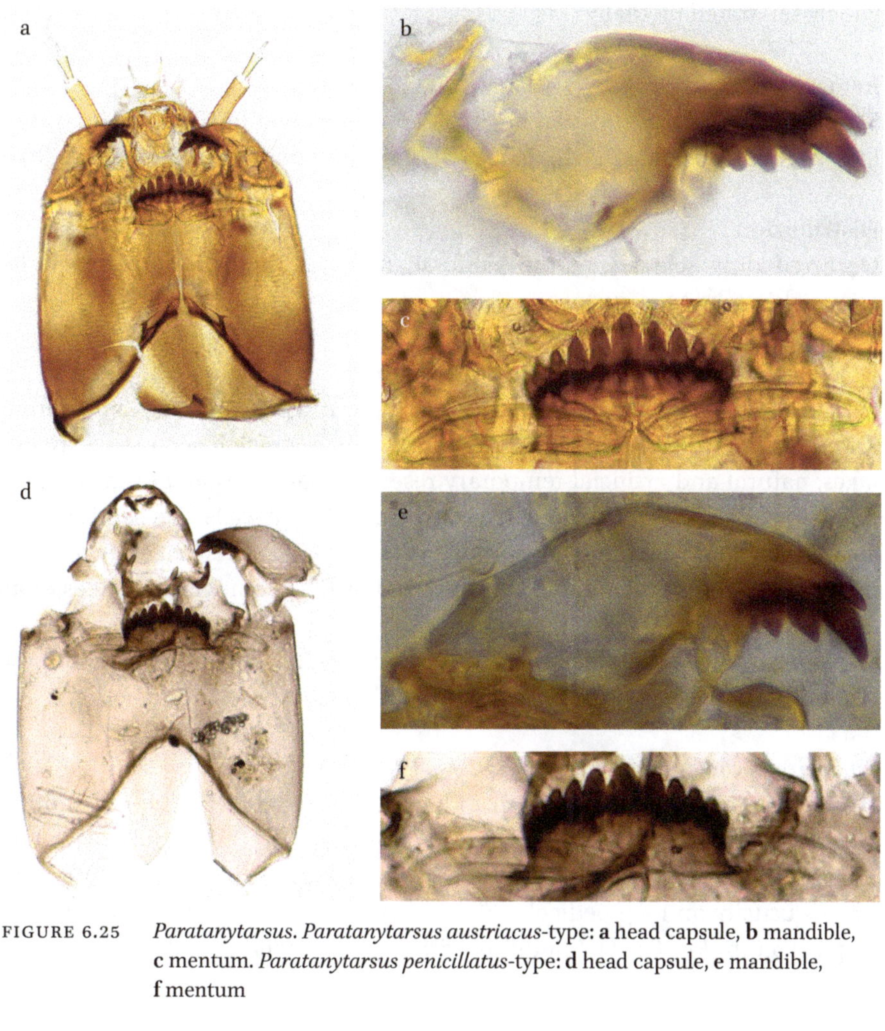

FIGURE 6.25 *Paratanytarsus. Paratanytarsus austriacus*-type: **a** head capsule, **b** mandible, **c** mentum. *Paratanytarsus penicillatus*-type: **d** head capsule, **e** mandible, **f** mentum

& Lazzara 2004, Samartin et al. 2012), while *P. penicillatus*-type prefers warmer waters (Brooks et al. 2007).

Stempellina Thienemann & Bause (Fig. 6.26)
Head capsule light brown, **occipital margin** brown. **Occipital plate** weakly developed (Fig. 6.26a). **Antenna** 5-segmented, placed on conspicuous pedestal bearing mesal palmate process, any additional spur absents. Lauterborn organs large, placed on short pedicels, occasionally extending slightly beyond flagellum (Fig. 6.26b). **Premandible** with 2–5 teeth. Dorsal **mandibular** tooth yellow, apical tooth and 2 inner teeth darker (Fig. 6.26d). **Mentum** with pale rounded or laterally slightly notched median tooth and 6 pairs of darker laterals, regularly decreasing in size (Fig. 6.26c). **Ventromental plates** fan-shaped, widely separated medially (Fig. 6.26c).

Remarks
Stempellina head capsules can be distinguished from other Tanytarsini by the large, palmate expansion which occurs on the antennal pedestal (Fig. 6.26b).

Distribution
Occurred only scarcely in the subfossil material in the reservoirs of the Štiavnické vrchy Mts. (SK).

Ecology and palaeoecological interpretation
Larvae of *Stempellina* are considered eurytopic and eurythermic, known from different types of lotic habitats, preferring slowly flowing sections of rivers, lakes, natural and artificial temporary reservoirs, marshes and even brackish habitats (Giłka 2005). They require coarse, grainy materials, like sand, to construct their long, curved transportable cases. The occurrence of *Stempellina* may be a response to suitable substrate rather than temperature (Rees et al. 2008).

Its occurrence in lake sediment records can indicate oligo- to mesotrophic conditions and oxygenated littoral with sandy and silty substrata.

Tanytarsus van der Wulp (Fig. 6.27)
Head capsules yellow to brownish with dark brown to black **occipital margin.** **Occipital plate** present, well developed (Figs. 6.27a,d,g,j). **Antenna** placed on tall pedestal (2–3 times as long as broad), usually without distal spur. Lauterborn organs usually on long pedicels (Fig. 6.27n). **Premandible** with 3 (usually) to 5 apical teeth (Fig. 6.27f). **Mandible** with 2–3 inner teeth, 1–2 dorsal teeth and

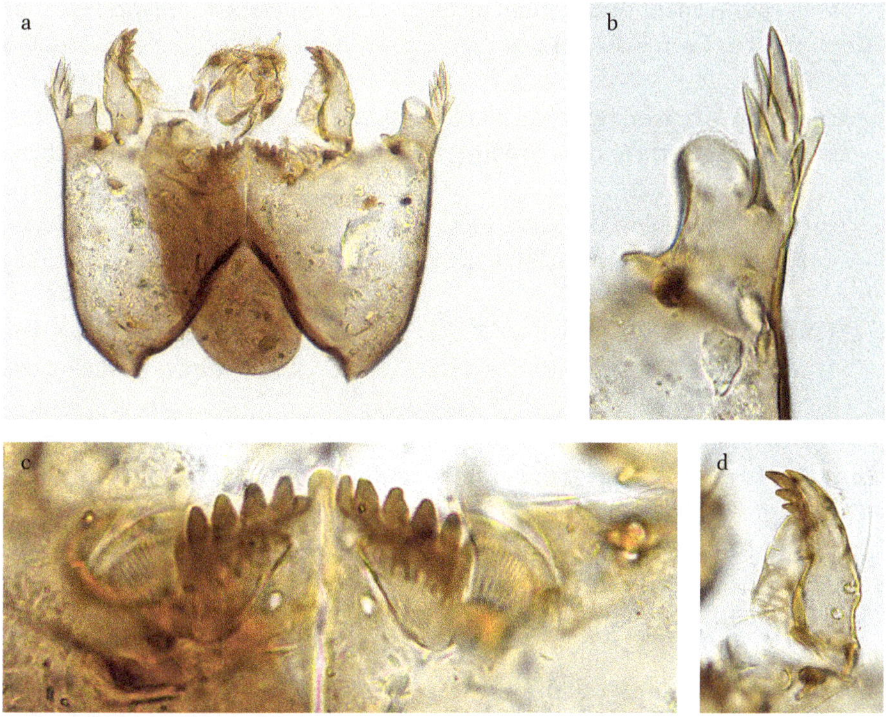

FIGURE 6.26 *Stempellina.* **a** head capsule, **b** antennal pedestal, **c** mentum, **d** mandible

in some species/morphotypes, 1–2 surficial teeth may be present additionally (Figs. 6.27c,f,i,l). **Mentum** with a single rounded or laterally notched median tooth and 5 pairs of lateral teeth (Figs. 6.27b,e,h,lk). **Ventromental plates** narrowly separated medially, long, narrow and fine striated (Figs. 6.27b,e,h,k).

Remarks

As mentioned above, there are considerable difficulties in distinguishing head capsules of *Tanytarsus* from *Micropsectra* (Fig. 6.24) and, to a lesser extent, from *Paratanytarsus* (Fig. 6.25) (see comments to *Micropsectra* and *Paratanytarsus*). Note that *Corynocera oliveri*-type with a large surficial tooth on the mandible is similar to *Tanytarsus lugens*-type. However, while the outer lateral teeth of the mentum are superimposed in *Corynocera*, in *T. lugens*-type should be clearly separated.

Morphotypes

Four morphotypes were distinguished following Brooks et al. (2007).

1 Small spur present on antennal pedestal .... *Tanytarsus pallidicornis*-type
– Spur absent from pedestal 2

2 Teeth of mentum on two focal plains: 3 middle teeth appear to be positioned above lateral teeth; mandible with 3 inner teeth, 2 dorsal teeth and 1 surficial tooth .... *Tanytarsus lugens*-type
– Teeth of mentum on the same focal plane; mandible with 2–3 inner teeth, 1 dorsal tooth and no surficial tooth 3

3 Mandible with 2 inner teeth .... *Tanytarsus nemorosus*-type
– Mandible with 3 inner teeth .... *Tanytarsus mendax*-type

***Tanytarsus lugens*-type**. Head capsule yellow with brown patch on submentum (Fig. 6.27a). Median tooth and first pair of lateral teeth are usually not at the same focal plane as the other laterals (Fig. 6.27b). Mandible with 3 inner teeth, 2 dorsal teeth and one surficial tooth, which is large and well visible (Fig. 6.27c). Antennal pedestals are long without spurs.
***Tanytarsus mendax*-type**. All teeth of the mentum at the same focal plane. Median tooth laterally notched, having a crown-like appearance (Fig. 6.27e). Mandible with 3 inner teeth and one dorsal tooth, surficial tooth absent (Fig. 6.27f). Antennal pedestal without spur. Occipital margin narrow, dark brown, post-occipital plate brown with slightly concave margin (Fig. 6.27d)
***Tanytarsus nemorosus*-type**. The type is similar to *T. mendax*-type but has only 2 inner teeth on mandible (Fig. 6.27i). Median tooth is lighter than lateral teeth and always rounded without lateral notches (Fig. 6.27h). Post-occipital plate narrow, only slightly darker than rest of head capsule (which is yellow) (Fig. 6.27g). According to the key in Brooks et al. (2007) this morphotype corresponds to *Tanytarsus nemorosus*-type. However, the book does not bring a photo of the morphotype, thus detailed comparison of our morphotype with the one described in Brooks et al. (2007) was not possible.
***Tanytarsus pallidicornis*-type**. All teeth of the mentum on the same focal plane (Fig. 6.27k). Antennal pedestals with short, rounded spurs (Fig. 6.27m). Mandible with 3 inner teeth (Fig. 6.27l). Occipital margin broader than in *Tanytarsus mendax*-type, dark brown to black, post-occipital plate brown with pale, straight margin (Fig. 6.27j).

Distribution

Generally, *Tanytarsus* remains were common in lake sediments, found in all lake districts. *Tanytarsus lugens*-type was one of the most frequent morphotypes in

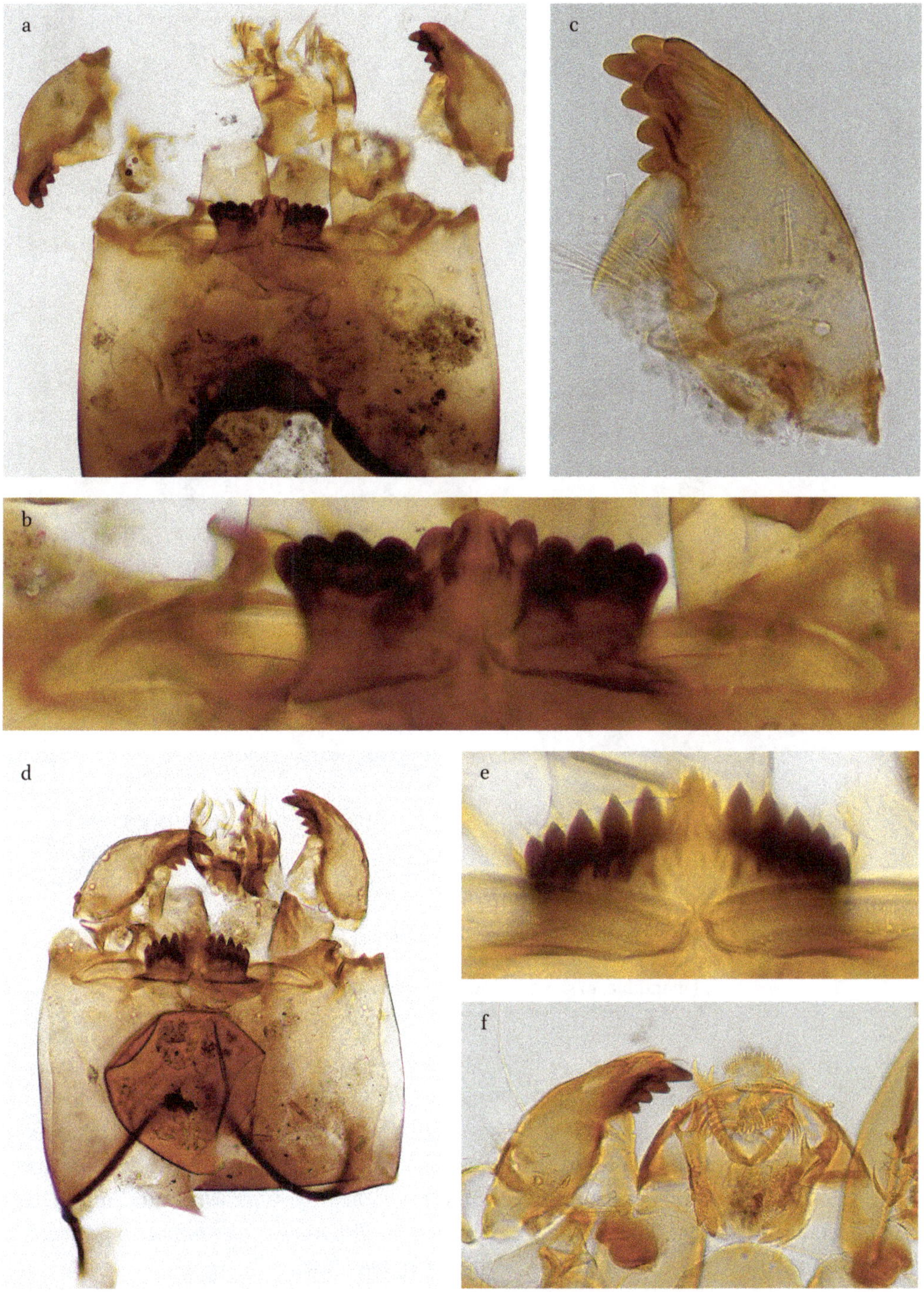

FIGURE 6.27 *Tanytarsus*. *Tanytarsus lugens*-type: **a** head capsule, **b** mentum, **c** mandible. *Tanytarsus mendax*-type: **d** head capsule, **e** mentum, **f** mandible and premandible

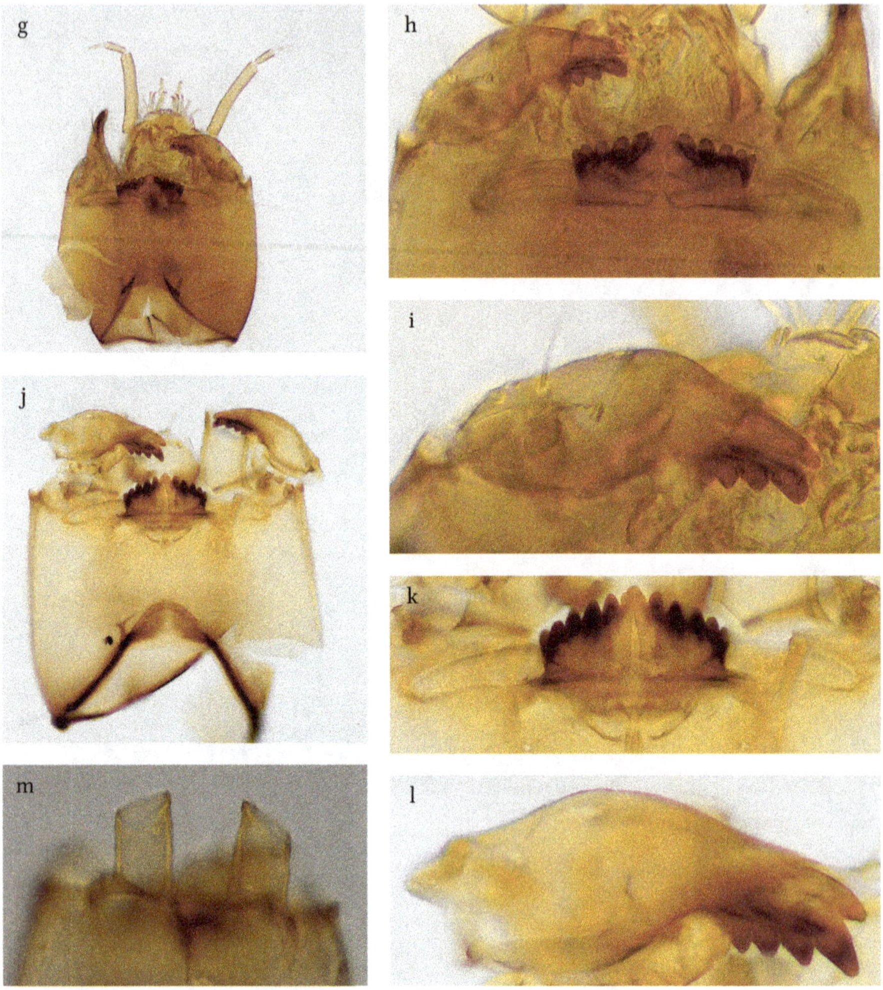

FIGURE 6.27 (continuation) *Tanytarsus nemorosus*-type: **g** head capsule, **h** mentum, **i** mandible. *Tanytarsus pallidicornis*-type: **j** head capsule, **k** mentum, **l** mandible, **m** short, rounded spurs on antennae pedestals

the Tatra Mts. lakes (SK, PL), usually absent in the coldest ones. *Tanytarsus mendax*-type was found relatively frequently in warmer lakes situated at lower altitudes in the Tatra Mts. (SK, PL), the Chornohora and Svydovets Massifs (UA), and the reservoirs in the Štiavnické vrchy Mts.; this morphotype seems to tolerate acidic conditions. *Tanytarsus pallidicornis*-type was found rarely in acidic Tatra Mts. lakes (SK, PL). Remains of *Tanytarsus nemorosus*-type were found only in the mesotrophic Morské oko lake (SK).

Ecology and palaeoecological interpretation
Tanytarsus is one of the most diverse and most widely distributed genera in Chironomidae (Lin et al. 2018). Larvae occur virtually in all types of freshwaters, some species are marine, and at least one species is terrestric. They are typical bottom dwellers, constructing usually long, soft tubes fixed to the substrate (Epler et al. 2013).

Tanytarsus remains often predominate (together with *Micropsectra*) in the subfossil material, so they should be considered carefully for palaeolimnological interpretation.

Tanytarsus lugens-type, commonly recorded in stratigraphic records from lakes in Northern Europe, is traditionally considered a cold-adapted taxon and an indicator of oligotrophy (Sæther 1979, Heiri & Lotter 2010). In fact, *Tanytarsus lugens* (Kieffer, 1916), a typical member of the "*lugens*" community (sensu Hofmann 1971), is a relatively cold-stenothermic species inhabiting the profundal of Central European lakes, while in Northern Europe the larvae may occur in the littoral (Hofmann 1971). However, the "*lugens*-type" mandible is present not only in *T. lugens* group, but in several otherwise morphologically distinct groups (Ekrem 2004, Cuppen & Tempelman 2018). For example, subfossil remains collected from lake sediments in the Tatra Mts. and identified as *lugens*-type actually belonged to *T. bathophilus* Kieffer, 1911, while *T. lugens* has not been found in the modern samples. *T. bathophilus* is not restricted to oligotrophic lakes and it can also be found in lakes with higher trophic status (Ekrem 2004). The species is widespread in the Tatra Mts., irrespective of elevation or lake morphology, but apparently avoids acidified lakes (Bitušík et al., 2006, Svitok et al. 2021). *Tanytarsus pallidicornis*-type is a warm-temperature indicator confined to lowland mesotrophic lakes (Laroque & Hall 2003, Samartin et al. 2012). *Tanytarsus mendax*-type appears to be eurytopic without clear relationship to key environmental variables (Langdon et al. 2010) most likely because this type includes several species with various ecological requirements. For example, *T. mendax*-type morphologically corresponds with *Tanytarsus gregarius*, which occurs mainly in acidic lakes (including dystrophic), but also in non-acidified lakes situated at lower elevations irrespective of their size or depth (Bitušík et al. 2006).

Tanytarsini indet. (Fig. 6.28)
Head capsule brown with slightly darker **occipital margin** (Fig. 6.28a). **Mandible** with robust, blunt apical tooth, 2 inner teeth and 1 dorsal tooth, all brown; obvious hump present on the outer margin (Fig. 6.28c). **Mentum** strongly arched, with a tall and slender median tooth and 5 lateral teeth, 1 to

4 subequal in size, outermost minute (Fig. 6.26b). **Ventromental plates** very close medially, slightly wider than width of mentum; they are also more fan shaped than typical Tanytarsini plates (Fig. 6.26b). Due to missing characters important for generic placement, this morphotype was not assigned to a genus, but its appearance indicates that it could be close to *Neozavrelia*, *Paratanytarsus* or *Sublettea*.

Morphotypes

One morphotype was recorded without placement into a genus.

Distribution

The morphotype was present only in one Tatra Mts. lake (PL).

Ecology and palaeoecological interpretation

The only remain was found in a large, deep (max. depth 76 m) subalpine lake (1580 m a.s.l.) with pH 7.14 (Czarny Staw pod Rysami).

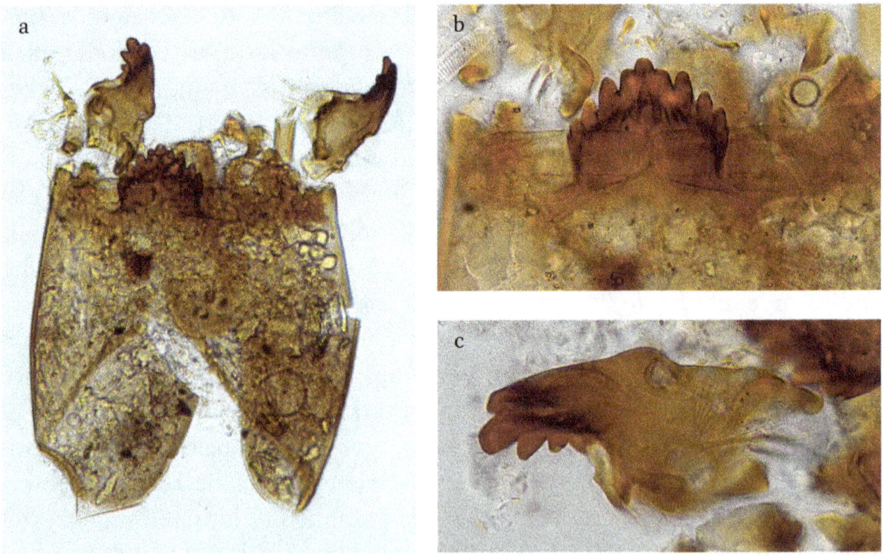

FIGURE 6.28 Tanytarsini indet. **a** head capsule, **b** mentum, **c** mandible

CHAPTER 7

Podonominae

The characteristic feature of Podonominae larvae is the absence of premandible and weak ventromental plates. The shape of mentum is unique and consists of a single median tooth and 7–15 pairs of lateral teeth. Outer margin of mandible is typically strongly bent in the middle, with a large apical, a small outer (so called subapical), and 4–9 inner teeth. The subfamily comprises two tribes, Podonomini and Boreochlini, both including six genera.

Parochlus **Enderlein** (Fig. 7.1)
Head capsule light brown, small to medium size. **Mentum** characteristic, comprises of single long median tooth and 7 lateral teeth, 3rd lateral tooth longer and more robust than 1st and 2nd laterals; outer lateral teeth gradually diminishing in size. **Ventromental plates** weakly developed. **Mandibles** with 1 dorsal tooth and 5 inner teeth; apical and 2nd inner tooth are the largest, 1st inner tooth is minute. **Premandible** absent.

Remarks
The unique shape of mentum and mandible of *Parochlus* will separate the genus from other chironomids.

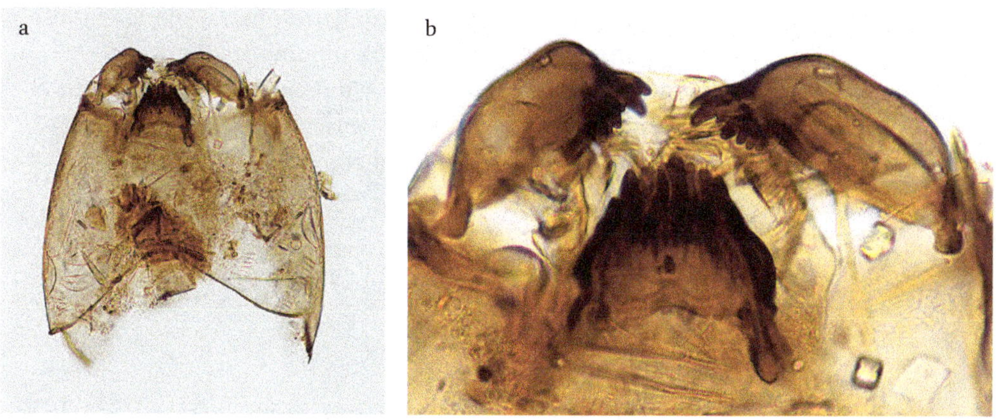

FIGURE 7.1 *Parochlus* sp. **a** head capsule, **b** mentum and mandibles

 | DOI:10.1163/9789004694781_008

Morphotypes

There is only species known from the Palaearctic, *Parochlus kiefferi* (Garrett, 1925), consequently the recorded morphotype most likely represents this species.

Distribution

One head capsule was found in a short sediment sequence taken from Vrbické pleso lake, Low Tatra Mts. (SK) corresponding to the Little Ice Age. This is the first record of the genus from Slovakia (Jambrović 2022).

Ecology and palaeolimnological interpretation

Parochlus is a species rich, cold-stenothermic genus mainly occurring in running waters of high latitudes and elevations. With 45 species, the genus shows wider distribution than any other genus of the subfamily (Brundin 1966). In Europe, however, it is represented by one species, *Parochlus kiefferi* (Garrett, 1925) that has been recorded in northern and central Europe as well as Ireland and the British Isles (Sæther and Spies 2004). Larvae of the species live among mosses in fast-flowing brooks, streams and rivers (Brundin 1983) from eukrenal to epirhitral (Moog 1995) but have also been recorded in high-altitude glacial ponds in the Italian Alps (Lencioni 2009). Larvae of *P. kiefferi* are grazers and gathering collectors (detritus feeders) (Moog 1995). Records of *P. kiefferi* emphasize the cold-stenothermal character of the species (see Lencioni et al. 2009). It is frequently found together with species limited to very cold conditions, such as *Pseudodiamesa nivosa* (Goetghebuer 1928) and *Micropsectra radialis* Goetghebuer, 1939 (Lencioni et al. 2009). Both species belong to those with the coldest temperature optima in the Tatra training-set.

Parochlus larval remains are very rare in lake sediments and usually occur in cold lakes of high latitudes (Brooks et al. 2007) and altitudes (Hamerlík et al. 2022, Jambrović 2022). Larval subfossils will indicate cold conditions.

CHAPTER 8

Other Insect Remains in Lake Sediments

Chironomidae represent the most abundant and diverse insect remains in lake sediments but by far not the only ones. In fact, nearly all animal groups leave some sort of identifiable morphological remnants in the sediments (Frey 1964). However, these ecologically important aquatic insect groups endure underused in palaeoenvironmental studies, partly because of the lack of identification keys and catalogues that would increase awareness and enable identification of subfossil remains (Courtney-Mustaphi et al. 2024). Indeed, compared to the more common animal remains, e.g., subfossil chironomids, for the identification of which have been published many keys covering different biogeographical regions (e.g., Eggermont & Verschuren 2004a,b, Brooks et al. 2007, Dieffenbacher-Krall et al. 2008, Hamerlík & Silva 2021), keys and manuals of other insect remains are rare. To our knowledge, the first comprehensive study dedicated to animal remains in Quaternary lakes and bog sediments was published 60 years ago by Frey (1964). Smol et al. (2001) provide detailed information on zoological indicators preserved in lake sediments including their taphonomy, ecology and main environmental drivers, methodological aspects of sampling and slide preparation, identification and taxonomy, and the use of different groups in tracking past environmental change. In a study analysing non-chironomid aquatic invertebrates (including a photography material of the animal remains) in a salinity gradient of western Uganda maar lakes, Rumes et al. (2005) identified different Dipteran families indicating saline environments. They also concluded that in small, simple lake basins, subfossil assemblages provide a more complete inventory of local aquatic invertebrate communities than an exploratory live sampling of littoral, profundal and pelagic habitats. Courtney-Mustaphi et al. (2024) provide photographs of both preidentified aquatic invertebrate specimens that have been chemically cleared to preserve structures that are observed in subfossil remains and examples of these structures from Quaternary lake-sediment samples. They demonstrate mandibles (and in some cases pygopodia and claw appendages) from several taxonomic groups including Crustacea (Amphipoda, Isopoda, Ostracoda, and Notostraca), Insecta (Coleoptera, Diptera, Ephemeroptera, Hemiptera, Odonata, Lepidoptera, Megaloptera, Plecoptera, and Trichoptera).

In this chapter, we give an overview of the most common remains of aquatic and terrestrial insects and their palaeolimnological implications. However, rather than being exhaustive, we aim at highlighting the great variability of

 | DOI:10.1163/9789004694781_009

other, often overlooked insect remains in lake sediments and encourage the reader to use them as complementary data. We included only insect groups where we (or our colleagues) had photography material, thus some groups, such as Hemiptera, Lepidoptera, Odonata, and many families of Diptera are missing. Thes groups are, however, well documented in Courtney-Mustaphi et al. (2024).

Diptera (flies)

Ceratopogonidae (biting midges, Fig. 8.1). Usually, the whole larval head capsule, which is small, prolonged, or broadly conical in shape (Figs. 8.1a,b), is preserved in the sediment. Even though the head capsule may remind chironomid taxa with small, prolonged heads, such as *Corynoneura* and *Thienemanniella*, the strong sclerotization and distinct shape of mentum and mandibles (Fig. 8.1c) can separate Ceratopogonidae. Aquatic larvae of this large family (comprising also many semi-terrestrial species) are mostly predators and live in shallow standing waters, normally at the margin; some species occupy streams. The higher proportion of ceratopogonid head capsules in a shallow lake on the Kola Peninsula, Russia, was interpreted as lake-level decline and an expansion of semiaquatic habitats along the lake shore (Ilyashuk et al. 2005). Significant changes in biting midge remains in the sediments of a lake in the Maramures Mts., Romania, most likely also indicate habitat changes and transition from lake to a peat bog (unpublished data). Szadziewski et al. (1997) can be used (with certain limitations) to identify subfossil material to genus level.

Chaoboridae (phantom midges, Fig. 8.2). In lake sediments, the characteristic remains of this family are the mandibles, consisting of three large teeth with a smaller tooth between the two larger ones (Fig. 8.2a, b). Occasionally, other larval mouthparts (Fig. 8.2c, d) and the pupal thoracic horn (Fig. 8.13f) can also be preserved. Larval *Chaoborus* typically inhabit the pelagic zone of standing waters, such as ponds and reservoirs, though some species (e.g., C. *pallidus*) are found exclusively in small forest lakes and temporary ponds. The transparent chaoborid larvae are planktonic and can regulate their vertical position in the water column (moving between the epilimnion, hypolimnion, and bottom sediments), which helps them avoid visual predators like fish. The larvae are predators that capture small planktonic prey from the water using their prehensile antennae (Walker 2001). Multiple studies suggested that the main driver of chaoborid assemblage composition is the fish presence/absence (e.g., Uutala 1990, Tolonen et al. 2012). On the other hand, some species are well adapted to co-exist with fish (such as *C. flavicans*) and their distribution is

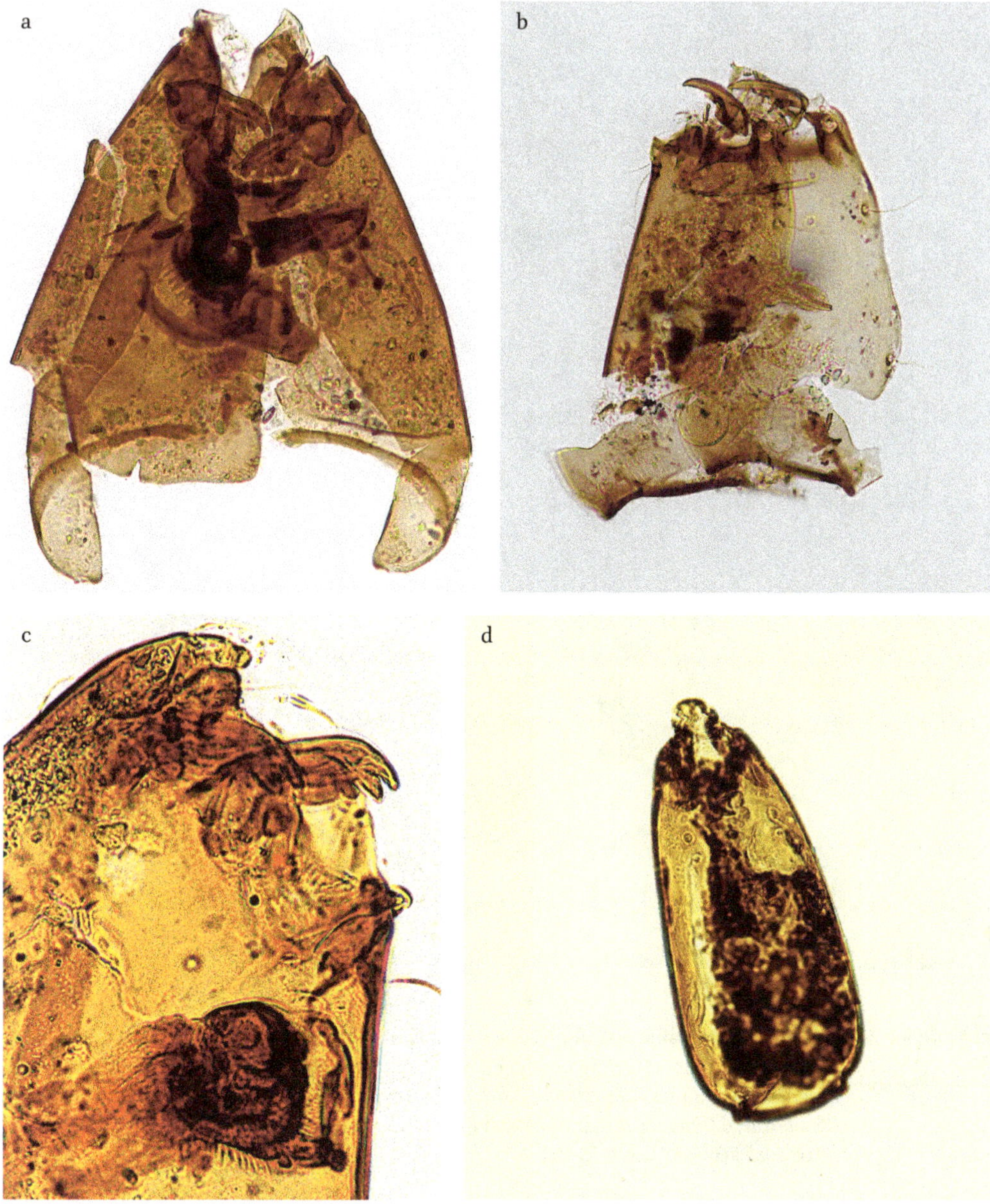

FIGURE 8.1 Ceratopogonidae head capsule remains as typically found in lake sediments. **a–c** Ceratopogonidae: *Dasyhelea*-type (**c** shows detail of head capsule with mandibles and mentum), **d** *Bezzia*-type head capsule
PHOTO D. VONDRÁK (D)

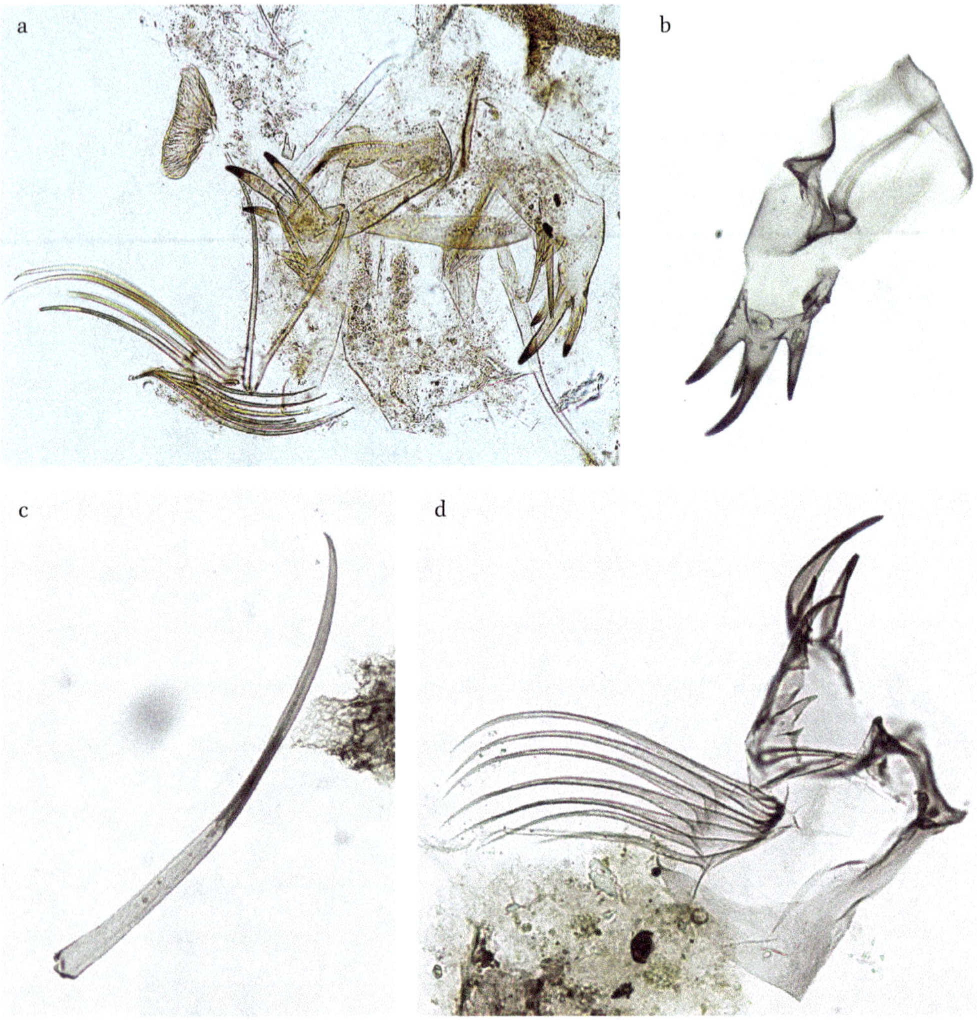

FIGURE 8.2 Chaoboridae remains. **a** Well-preserved *Chaoborus* head capsule showing both mandibles and labral filaments (*C. flavicans*-type), **b** part of *Chaoborus* mandible as usually preserved in sediment samples (*C. crystallinus/obscuripes*-type), **c** most likely antennal seta of *Chaoborus*, **d** *Chaoborus* mandible and mandibular fan
PHOTO A. FORMÁČKOVÁ (B–D)

mainly governed by the interactive effects of fish abundance, water transparency and lake morphometry (presence and extent of oxygen-poor hypolimnetic refuge). Chaoborid subfossils were also used to infer past hypolimnetic oxygen (Sweetman & Smol 2006, Quinlan & Smol 2010), water-level fluctuations and temperature in small shallow lakes (Luoto & Nevalainen 2009). Due to different relationships with fish populations, identification of different *Chaoborus*

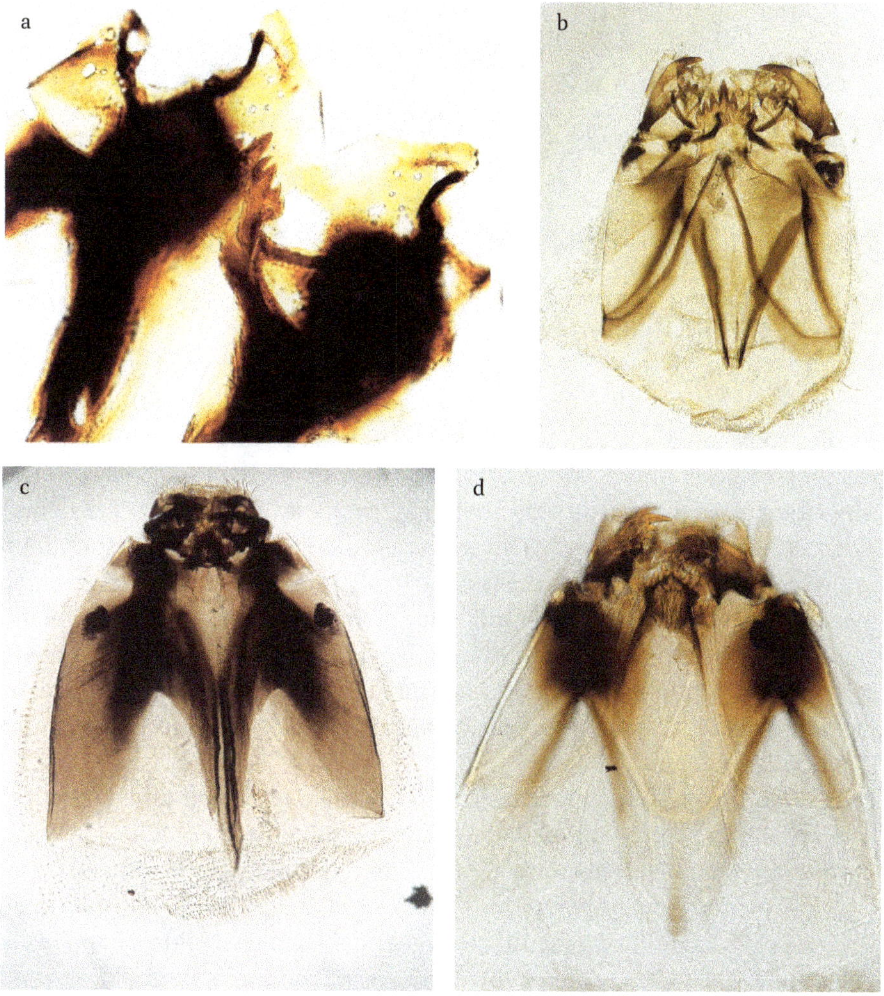

FIGURE 8.3 Limoniidae head capsules. **a** subfossil remain, **b–d** recent head capsules showing different color patterns
PHOTO D. VONDRÁK (A)

species may be important for reliable palaeolimnological interpretation. For instance, larger larvae of *Chaoborus obscuripes* (van der Wulp, 1859) with darker pigmentation are more sensitive to visually dependent predators (Stenson 1981) and thus will indicate fishless conditions in small, shallow, nutrient-poor, meso- and polyhumic ponds. Uutala (1990) and Salmela et al. (2021) provide identification keys for larval chaoborids.

Limoniidae (craneflies, Fig. 8.3). Remains, usually represented by the whole head capsule, are rare in lake sediments. Larvae of this large family are usually

FIGURE 8.4 Pediciidae head capsule remain
PHOTO D. VONDRÁK

aquatic or semi-aquatic. Limoniidae larvae occupy a wide range of habitats, such as soils and decaying plant materials in swamps and marshes, soils and wet mosses along stream borders, fast flowing streams, intertidal zones, brackish waters as well as madicolous habitats (i.e., constant wet film on stones) along streams and rivers (Špaček et al. 2019). The ecological information gained from the remains may thus not be straightforward; however, in general, they may indicate flowing water influence or erosion events in the catchment. Species and genus identification is challenging even when living larvae are available. Mauch (2017) provides a key for recent larvae of European genera, including detailed drawings of head capsules, which can also be useful for identifying subfossils.

Pediciidae (hairy-eyed craneflies, Fig. 8.4). A relatively small dipteran family that was erected from Limoniidae. Larvae are primarily rheophilic and occur in various types of running waters from springs (rheocrenes) to estuaries, usually not tolerating toxic pollution (Špaček et al. 2019). Rozkošný (1980) and Ujvárosi et al. (2010) bring identification keys of larval Pediciidae. Remains in the sediment sequence would indicate increased running water influence.

Psychodidae (moth flies, Fig. 8.5). Larval head capsules (Figs. 8.5a,b,c) and occasionally parts of the pupa (e.g., thoracic horn, Fig. 8.5d) can be preserved in lake sediments. Larvae of this big family live in aquatic to semi-terrestrial habitats but species vary considerably in their ecology. They can be found in fast flowing streams and rivers, hygropetric habitats, and dendrotelmata; larvae of some species can inhabit plumbing drains where they feed on microorganisms. Larvae of the genus *Psychoda* are common in settling ponds and trickling filters of sewage treatment facilities. In general, increase of Psychodidae in streams may indicate sewage contamination (especially those from the Psychodini group), but it is important to note that some groups (i.e., Pericomini) will be more abundant in uncontaminated waters (Masteller & Wagner 1984). Thus, Psychodidae remains may indicate both increased inflow

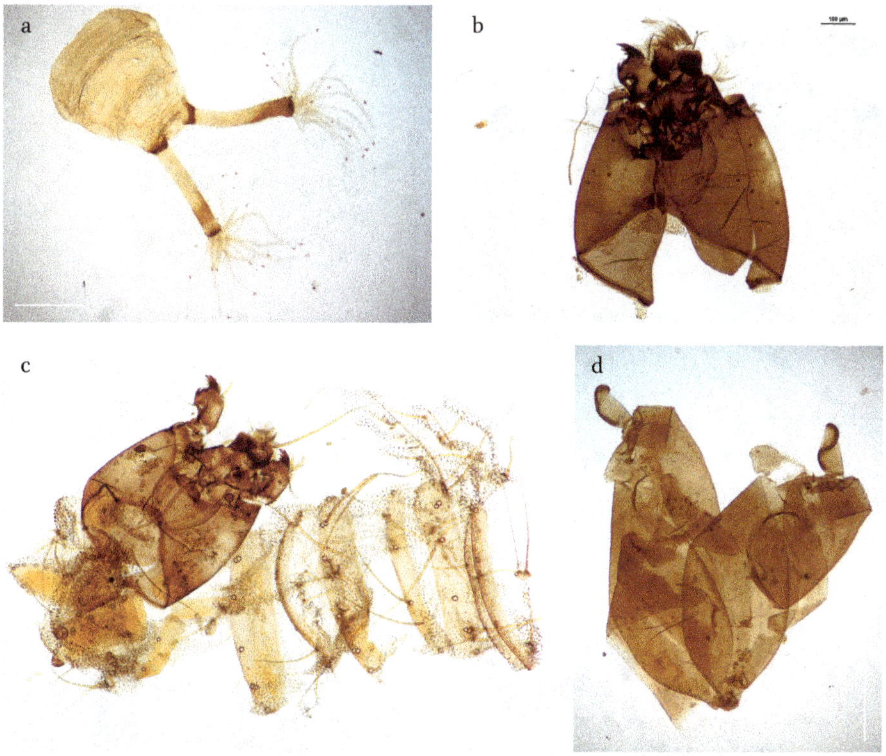

FIGURE 8.5 Psychodidae larval and pupal remains. **a** and **b** head capsule, **c** head capsule and larval exuvium, **d** fragment of pupa with thoracic horns

and heavy organic pollution and reaching higher taxonomic resolution (genus level) may be required to distinguish between them.

Simuliidae (black flies, Fig. 8.6). Relatively large, usually heavily sclerotized, dark larval head capsules (Figs. 8.6a,b) can be encountered in the sediment record. The unique shape of hypostomal teeth (a structure similar but not homologous to mentum of chironomids, Figs. 8.6c–f) makes confusion with other dipterans unlikely. In case the whole head capsule is preserved, the shape of the postgenal cleft can be useful for identification, too. This large family with world-wide distribution is notorious for the blood-sucking habit of the adult females and being vectors for diseases. Black fly larvae are found attached to hard surfaces exclusively in fast-flowing waters, where they inhabit diverse habitats from springs to large rivers. They can be abundant in plankton-rich waters immediately downstream of lakes, likely due to the combination of the plenitude of food and the low concentration of suspended mineral matter. Most species feed on suspended food particles (detritus, algae, and small animals)

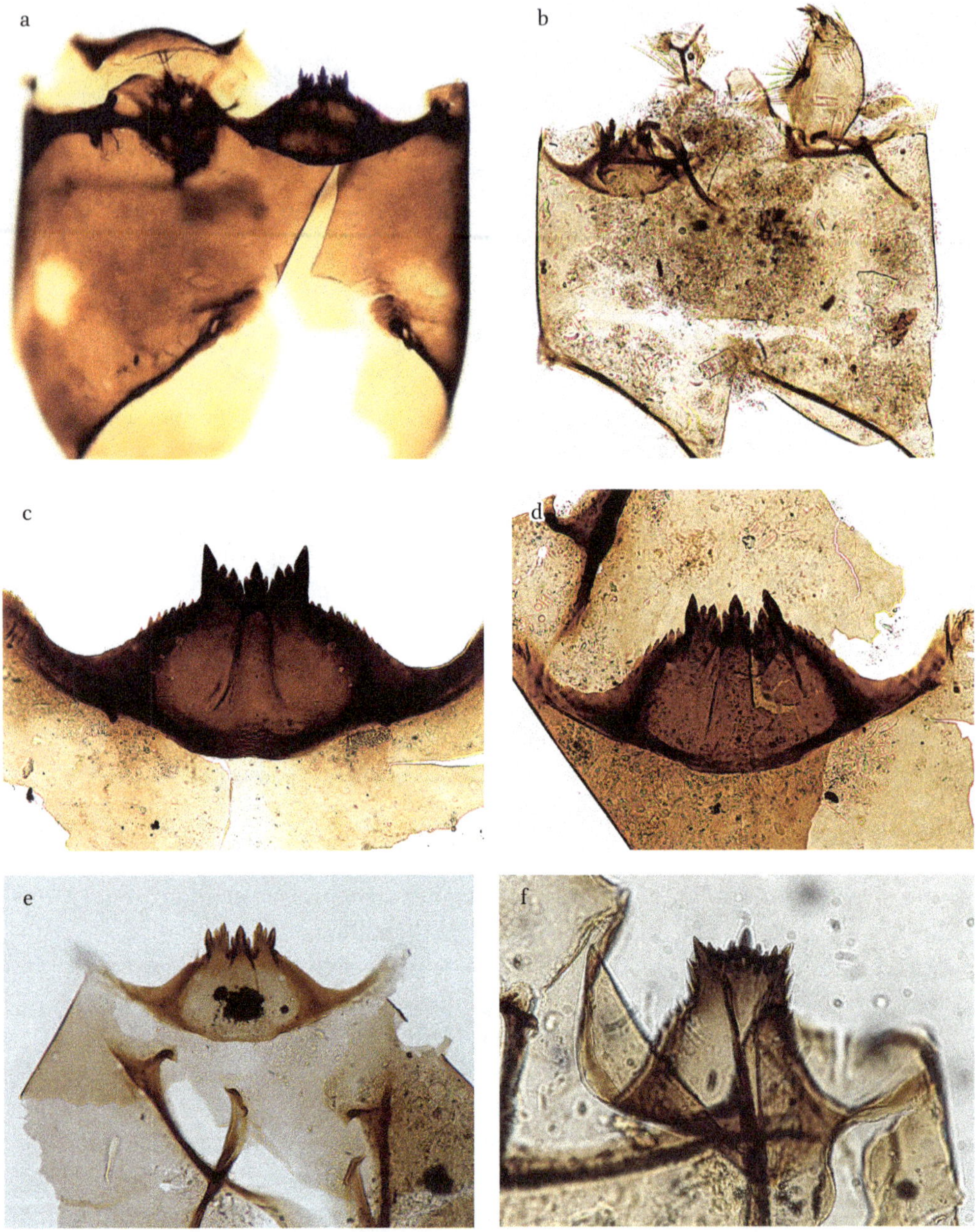

FIGURE 8.6 Simuliidae head capsules (**a**, **b**) and hypostoma (**c–f**). **c** *Prosimulium latimucro*, **d** *Prosimulium hirtipes/rufipes*, **e** Prosimuliini indet. **f** *Simulium* (*Nevermannia*) *vernum* group
FOTO: D. VONDRÁK (A), J. PETERSEN (F)

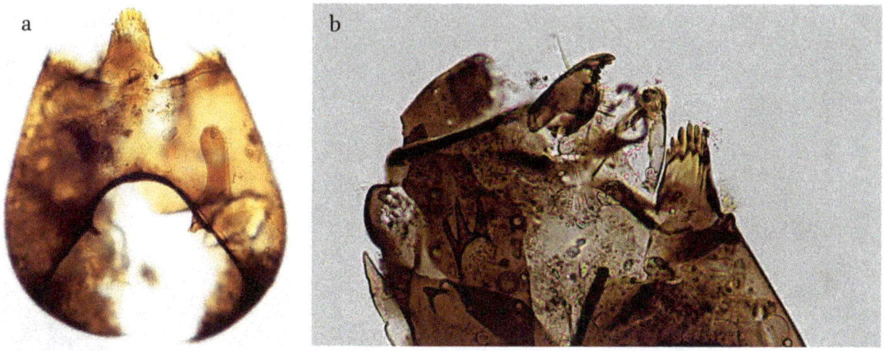

FIGURE 8.7 Thaumaleidae head capsule (**a**) and detail of showing mentum and mandible (**b**)
PHOTO D. VONDRÁK (A)

that they filter from the water current using a pair of labral fans (Walker 2001). Food quality and quantity, water temperature, and hydrodynamic characteristics drive the distribution of black fly larvae (Ross & Merritt 1987). Since larvae are confined to running waters, their subfossils have a potential as indicators of past hydrologic conditions and in turn, can provide useful information on palaeoclimatic reconstructions (Currie & Walker 1992). Currie & Walker (1992) provide a key to subfossil larvae of Nearctic genera based on features on hypostoma. Mauch (2017) brings a key to recent Simuliidae larvae including many detailed drawings of head capsules and mouthparts.

Thaumaleidae (solitary midges, Fig. 8.7). The head capsule (and entire larval body) closely resembles chironomids, but the numerous tubercles on the dorsal head (what also appears in a lower extent in the chironomid genus *Boreoheptagyia*) and the characteristic shape of mentum will distinguish thaumaleids (Figs. 8.7a,b). Larvae of this small family are semiaquatic, mainly inhabiting hygropetric habitats, such as wet rock walls and spray zones of waterfalls and riffles where they graze on diatoms. In streams they usually occupy the transitional zone between water and land; they can be numerous in concrete culvert pipes. Larvae prefer unpolluted natural waterbodies; therefore, they occur primarily in mountain areas, occasionally in the lowlands (Špaček et al. 2019). In the sediment sequence, thaumaleid subfossils may indicate cold inlet stream flow and the occurrence of hygropetric habitats.

Coleoptera (beetles, Fig. 8.8)

Because of their highly resistant exoskeleton, subfossil remains of beetles are often abundant in Quaternary deposits (Elias 2010). The main exoskeleton parts preserved in lake sediments are the mandibles, pronotum and the

FIGURE 8.8 Coleoptera remains. **a** elytrum of an adult beetle, **b** larval mandible
PHOTO D. VONDRÁK (A), A. FORMÁČKOVÁ (B)

elytra (a modified, hardened forewing, Fig. 8.8a). Recent beetle species seem to have been morphologically unchanged for several million years and thus are assumed to have the same ecological tolerances/preferences, which makes them well suited for palaeoecological studies (Elias 2010). Coleoptera is one of the largest orders of insects with hundreds of thousands of species described worldwide. Beetles occupy virtually every terrestrial and freshwater habitat on the planet, but many species are restricted to a specific habitat or operate within a narrow range of environmental conditions. The size of the adults varies from small (0.4 mm) to very large (200 mm) and most are usually heavily sclerotized and compact. Most beetles are herbivores, fungivores or carnivores and most adults can fly (Resh & Cardé 2009).

Moreover, many species are restricted to specific habitats and thermal conditions and thus they can serve as useful proxies when reconstructing past environments and climate change (Elias 2010).

Ephemeroptera (mayflies, Fig. 8.9)

Mayflies, also known as dayflies, shadflies, or even fishflies are the most primitive and ancient of the extant insect groups. The mandibles (Fig. 8.9a–c) of this large group of primarily aquatic insects can be encountered in lake sediments. Although some species can be found in ponds and lake littoral (e.g., *Cloeon*, *Caenis*, *Ameletus*), mayfly larvae primarily occupy running waters. Since larvae mostly reflect stream conditions, their remains can be used in palaeohydrological studies to reconstruct past changes in stream flow. Luoto et al. (2013) showed that the surface sediment accumulation of mayfly (and caddisfly)

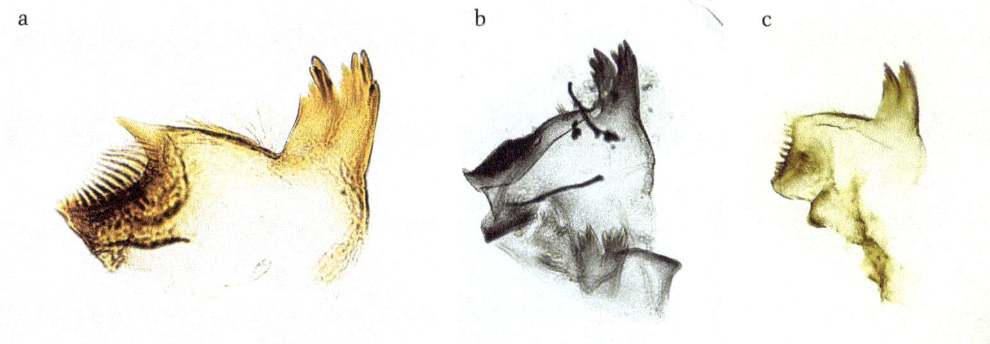

FIGURE 8.9 Ephemeroptera mandibles (**a–c**)
PHOTO D. VONDRÁK (A), A. FORMÁČKOVÁ (B–C)

mandibles was strongly related to stream flow intensity. Analyzing a sediment core from a Finnish lake they found that mandible accumulation reflects the amount of winter precipitation (i.e., spring floods caused by snow melt). Thus, increased occurrence of mayfly remains may be connected to water level fluctuations or increased inlet activity. Bauernfeind & Humpesch (2001) provide a key to recent mayfly larvae in central Europe. Courtney-Mustaphi et al. (2024) show a wide variety of ephemeropteran mandibles of different species/genera.

Megaloptera (alderflies, Fig. 8.10)
The most frequent remains of Megaloptera larvae are the powerful, strongly sclerotized mandibles and occasionally the frontoclypeus (Figs. 8.10a,b). Larvae of the Sialidae, the only family present in Europe, are predators with slender, pincer-like mandibles adapted for biting and holding prey (Fig. 8.10a). Sialidae larvae mainly live in soft sediments of standing waters but also occur in running waters. Mature larvae leave the water to pupate on land and after mating the adults place the egg mass on vegetation near water; after hatching the larvae immediately move to water. In shallow lakes, e.g., in the mountain lakes in the Ukrainian Carpathians, larval remains can be relatively abundant (pers. observation). Without knowing the exact species, palaeolimnological interpretation can be challenging, since larvae of the most common species live in different environments. While *Sialis lutaria* (Linné 1758) occupies mainly soft sediments of standing waters, usually lakes and ponds, and occasionally slow streams; *S. fuliginosa* Pictet 1836 is known from faster streams and rivers (Rozkošný et al. 1980). Meinander (1996) provides an identification key for the Sialidae larvae of Europe.

a b

FIGURE 8.10 Megaloptera larval remains. **a** *Sialis* mandible, and **b** *Sialis* frontoclypeus
PHOTO D. VONDRÁK (A,B)

Plecoptera (stoneflies, Fig. 8.11)
Stoneflies are a small ancient insect order whose larval remains, usually represented by the robust mandibles (Fig. 8.11a–e), have been rarely reported from lake sediments (Courtney-Mustaphi et al. 2024). The aquatic stages of stoneflies, also known as nymphs, feed on decaying leaves and wood, encrusting algae, or on other invertebrates; some species undergo ontogenetic diet shifts (DeWalt & Ower 2019 and references therein). The highest species richness of stoneflies occurs in cool mountains streams of temperate latitudes, but nymphs dwell in all stream sizes, inhabit high latitude or elevation lakes, tolerate a wide range of thermal regimes, and can complete their life cycles under a wide range of stream permanence conditions (DeWalt et al. 2015). The preferable substrates are rocks, bryophytes and litter. In general, Plecoptera nymphs require high dissolved oxygen concentrations and thus are very sensitive to environmental pollution (with few exceptions) and are often used in freshwater monitoring programs. Adult stoneflies are weak flyers and are found in riparian vegetation.

Stonefly remains in the sediment record will indicate intense cold, fast-flowing inlet and/or stony littoral conditions in a cold, oxygen-rich lake. When identifying/counting Plecoptera mandibles, it should be considered that the number of teeth and the form of the molar region is asymmetrical between the right and left mandibles (Courtney-Mustaphi et al. 2023). Teslenko & Zhiltzova (2009) present an identification key for adults and nymphs of Russia and adjacent countries.

Trichoptera (caddisflies, Fig. 8.12)
The main body parts of larval caddisflies preserved in lake sediments are the frontoclypeus (Fig. 8.12a–c), pygopodium (Fig. 8.12d), pronotum, mesonotum,

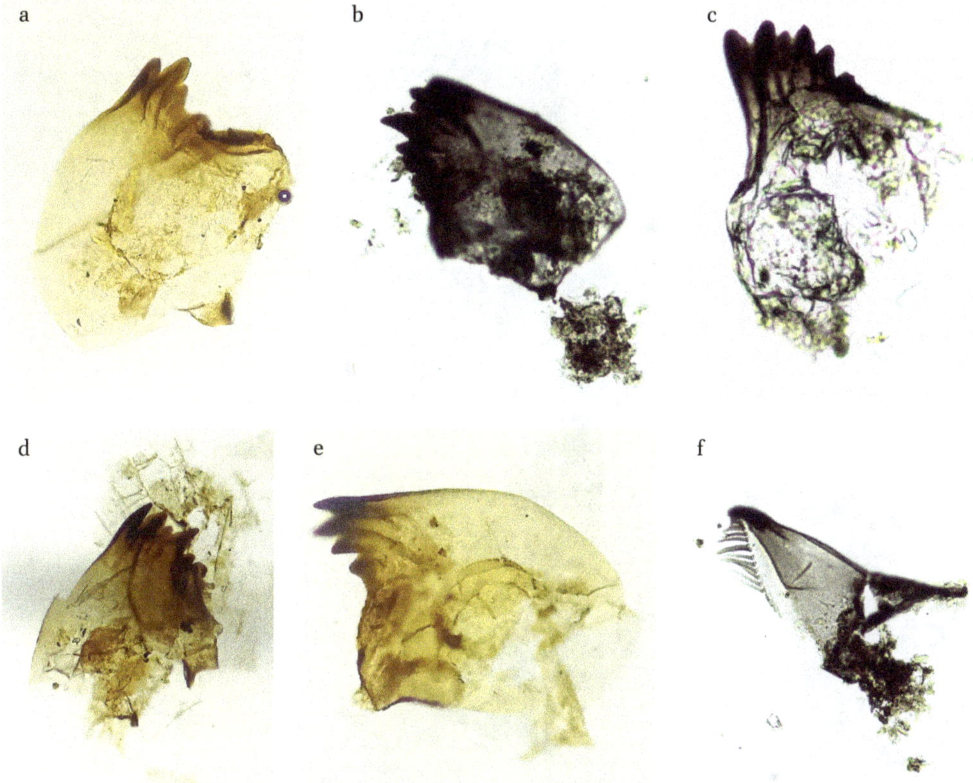

FIGURE 8.11 Plecoptera mandible remains (**a–e**) and maxilla (**f**) of different families
PHOTO A. FORMÁČKOVÁ (A–F)

and the larval case. Other fragments such as disarticulated leg segments, labrum and mandibles mandible (8.12e,f) can also be abundant, but these structures are less reliable for taxonomic identification. Caddisflies are one of the most diverse aquatic insect orders with more than 14 thousand species worldwide (Holzenthal et al. 2011). They are adapted to a wide range of habitats and exhibit many different feeding strategies including grazers/scrapers, filter feeders, shredders, deposit feeders and predators. Due to the different tolerance of species to environmental conditions, caddisfly larvae have been commonly used as indicators for ecological quality assessment of streams and rivers (Rosenberg & Resh 1993). Caddisflies larvae are (with few exceptions) aquatic and can be found in almost every type of freshwater habitat. Most species construct portable cases from different kinds of material, including plant fragments, mineral particles and mollusc shells, while other species construct fixed retreats on solid objects, often associated with silky nets for capturing

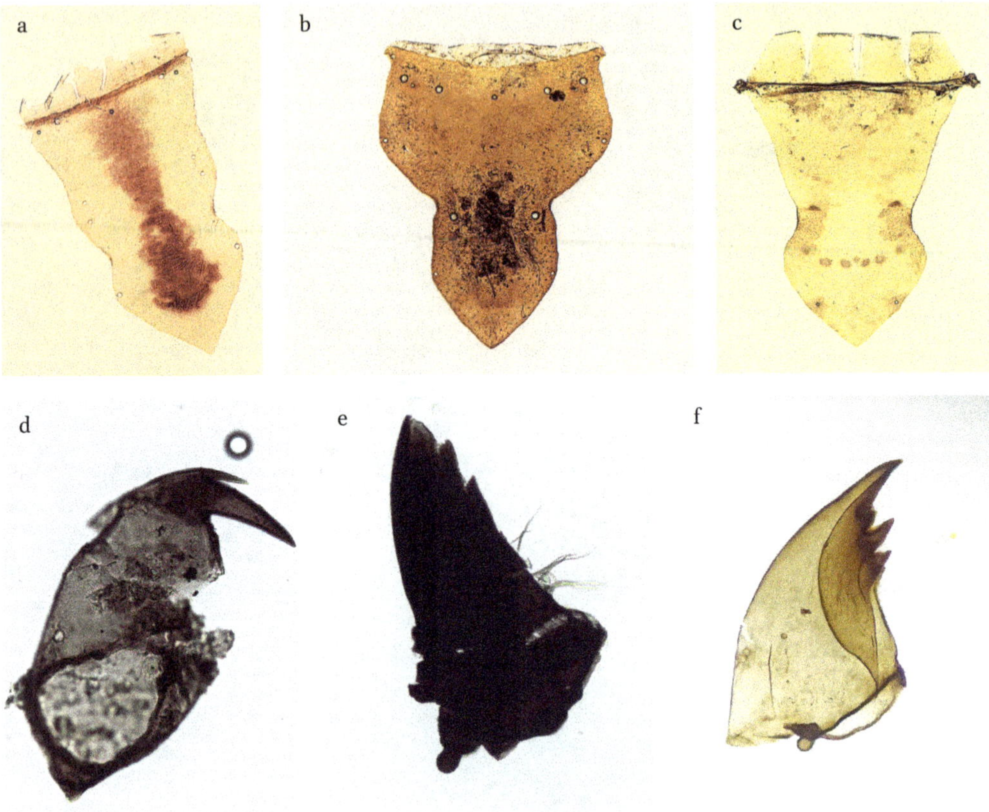

FIGURE 8.12 Trichoptera larval remains: frontoclypeus (**a–c**), pygopodium (**d**) and mandible (**e, f**). **a** Phryganeidae frontoclypeus, **b** Limnephilidae frontoclypeus, **c** Polycentropodidae frontoclypeus, **d** Limnephilidae pygopodium, **e** Limnephilidae mandible, **f** Polycentropodidae (?) mandible
PHOTO D. VONDRÁK (A–C), A. FORMÁČKOVÁ (D–F)

food, and a few species are free-living and only make a protective case prior to pupation (Wiggins 2004). Most free-living caddisflies are associated with running waters. Adults are aerial (with a few exceptions) but they usually have a very restricted range of dispersal. Eggs are embedded in a gelatinous substance and laid directly into the water, on rocks or on overhanging vegetation (Resh & Cardé 2009).

Caddisflies are well suited for palaeoecological studies because the recent species seem to have been morphologically unchanged for hundreds of thousands of years and therefore are expected to show the same ecological tolerances/preferences (Williams 1989). Caddisflies can be used to reconstruct climate and the local habitat by giving information about, e.g., temperature, the type of water body, water depth, flow regime, substrate and vegetation. Vondrák et al. (2019) performed a high-resolution analysis of the postglacial succession of a Czech lake of glacial origin to reconstruct its more than 11 000-year environmental history. While the community confirmed the acid, dystrophic state of the lake since its inception, it failed to detect any signs of anthropogenic acidification during the twentieth century. Luoto et al. (2013) used subfossil remains of caddisflies and mayflies to reconstruct past palaeohydrological changes. Waringer & Graf (2011) can be used for identification of recent European species.

Thoracic horns of dipteran pupae (Fig. 8.13)
Most dipteran pupae possess paired respiratory organs, known as thoracic horns. Pupae of some species, particularly of the free-living ones (e.g., Culicidae, Chaoboridae and subfamilies Tanypodinae and Podonominae of the Chironomidae) utilize atmospheric air and are equipped with exceptionally well-developed thoracic horns that can be steadily preserved in lake sediments. Owing to their unique shape and size, thoracic horns have diagnostic significance (sometimes allowing even genera/species identification) and thus may provide important supplementary information for palaeoreconstructions. Fig. 8.13 presents thoracic horns of groups with potential occurrence in lake sediments, such as Chironomidae (subfamily Tanypodinae, Fig. 8.13a–e), Chaoboridae (Fig. 8.13f), and Psychodidae (Fig. 8.13g).

FIGURE 8.13 Thoracic horns of some dipteran pupae. Chironomidae (subfamily Tanypodinae): **a** *Zavrelimyia barbatipes*, **b** *Macropelopia notata*, **c** *Ablabesmyia* sp., **d** *Procladius* cf. *choreus*, **e** *Procladius* sp. Chaoboridae: **f** *Chaoborus* sp. **g** Psychodidae indet. Figures a–c show horns of recent pupal exuviae, d–f are subfossil remains. Figures a–c were modified after Bitušík and Hamerlík (2014), Photo A. Formáčková (d, e)

Appendix 1

List of morphotypes and their distribution in surface sediments of the surveyed lakes. Numbers refer to the number of lakes in which the morphotype was recorded. SK – Slovakia, PL – Poland, UA – Ukraine, RO – Romania; Lt – Low Tatra Mountains, Ta – Tatra Mountains, Vi – Vihorlatské vrchy Mountains, St – Štiavnické vrchy Mountains, Uc – Ukrainian Carpathians, Ma – Maramures. Morphotypes marked with * were not recorded in the surface sediments but deeper in the sediment cores.

Country	SK				PL	UA	RO
Lake district	**Lt**	**Ta**	**Vi**	**St**	**Ta**	**Uc**	**Ma**
Podonominae							
*Parochlus**	1	–	–	–	–	–	–
Tanypodinae							
Ablabesmyia	1	2	1	3	6	1	–
*Derotanypus**	1	1	–	–	–	–	–
Macropelopia nebulosa-type	–	9	–	2	1	3	–
Procladius	1	26	1	3	12	10	1
Zavrelimyia	1	10	–	–	8	2	–
Diamesinae							
Diamesa bertrami-type	1	8	–	–	1	1	–
Diamesa zernyi/cinerella-type	1	11	–	–	7	3	–
Pseudodiamesa branicki-type	–	10	–	–	5	1	–
Pseudodiamesa nivosa-type	–	14	–	–	1	–	–
Pseudokiefferiella parva	–	5	–	–	1	–	–
Prodiamesinae							
Prodiamesa olivacea-type	1	3	–	3	2	1	–
Orthocladiinae							
Brillia bifida-type	1	2	–	2	–	–	–
Chaetocladius piger-type	1	6	–	1	5	–	–
Corynoneura arctica-type	1	11	–	–	10	3	–
Corynoneura edwardsi-type	–	1	–	–	4	1	–
Corynoneura lobata-type	1	–	–	–	1	2	–
Cricotopus (Cricotopus) cylindraceus-type	1	–	–	1	–	–	–
Cricotopus (Cricotopus) tremulus-type	–	6	–	1	8	–	–
Cricotopus (Isocladius) intersectus-type	1	–	–	2	1	2	–

 | DOI:10.1163/9789004694781_010

(*cont.*)

Country		SK			PL	UA	RO
Lake district	**Lt**	**Ta**	**Vi**	**St**	**Ta**	**Uc**	**Ma**
Cricotopus (*Isocladius*) *obnixus*-type	1	–	–	1	–	1	–
Cricotopus (*Isocladius*) *sylvestris*-type	–	5	–	–	4	1	–
Cricotopus (*Paratrichocladius*) *skirwithensis*-type	1	4	–	2	3	1	–
Cricotopus type Tatra	–	1	–	–	–	–	–
Diplocladius cultriger	–	1	–	–	–	1	–
Epoicocladius	–	–	1	–	–	–	–
Eukiefferiella brevicalcar-type	–	–	–	–	1	1	–
Eukiefferiella claripennis-type B	–	1	–	–	3	–	–
Eukiefferiella devonica-type	1	–	–	–	1	–	–
Eukiefferiella fittkaui-type	–	6	–	–	7	3	–
Gymnometriocnemus/*Bryophaenocladius*	–	12	–	–	3	1	–
Heterotrissocladius marcidus-type	1	32	–	–	12	1	–
Hydrobaenus conformis-type	–	1	–	–	–	–	–
Limnophyes/*Paralimnophyes*	1	5	–	3	5	9	1
Metriocnemus eurynotus-type	1	3	–	1	–	4	–
Metriocnemus fuscipes-type	1	2	–	–	1	1	–
Nanocladius rectinervis-type	–	–	–	1	–	–	–
Orthocladius (*Orthocladius*) *dentifer*-type	–	1	–	–	–	–	–
Orthocladius (*Mesorthocladius*) *frigidus*	–	3	–	–	–	–	–
Orthocladius (*Euorthocladius*) sp.	1	–	–	–	1	–	–
Orthocladius (*Symposiocladius*) *lignicola**	1	–	–	–	–	–	–
Paracladius	–	–	–	1	–	–	–
Parakiefferiella bathophila-type	1	–	–	3	–	–	–
Parametriocnemus/*Paraphaenocladius*	–	3	1	1	1	1	–
Paratrissocladius	–	–	–	1	–	–	–
Parorthocladius	–	2	–	–	1	–	–
Psectrocladius (*Allopsectrocladius*) *flavus*-type	–	2	–	1	–	1	–
Psectrocladius (*Mesopsectrocladius*) *barbatipes*-type	–	1	–	–	–	2	–
Psectrocladius (*Psectrocladius*) *psilopterus*-type	1	4	–	–	1	3	–

(*cont.*)

Country		SK			PL	UA	RO
Lake district	**Lt**	**Ta**	**Vi**	**St**	**Ta**	**Uc**	**Ma**
Psectrocladius (*Psectrocladius*) *sordidellus*-type	1	9	1	–	10	3	–
Pseudosmittia	1	–	–	–	1	2	–
Rheocricotopus effusus-type	1	–	–	2	–	1	–
Rheocricotopus fuscipes-type	–	–	–	1	–	1	–
Smittia/Parasmittia	1	3	1	3	3	3	1
Synorthocladius semivirens	–	1	–	2	1	–	–
Thienemanniella clavicornis-type	–	2	–	–	4	1	–
Tvetenia bavarica-type	–	1	–	3	3	–	–
Zalutschia mucronata-type	–	–	–	–	1	–	–
Zalutschia type B	–	4	–	–	–	–	–
Chironominae – Chironomini							
Benthalia	–	–	1	–	–	–	–
Chironomus anthracinus-type	1	5	–	3	2	6	1
Chironomus plumosus-type	1	2	1	3	2	7	1
Cladopelma laccophila-type	–	–	–	1	–	–	–
Cladopelma lateralis-type	–	1	1	1	1	–	–
Cryptochironomus	–	–	–	3	–	–	–
Cryptotendipes	–	–	–	2	–	–	–
Dicrotendipes lobiger	1	–	–	–	–	–	–
Dicrotendipes nervosus-type	–	–	1	3	1	–	–
Dicrotendipes notatus-type	1	–	1	3	3	–	–
Endochironomus albipennis-type	–	–	–	3	–	–	–
Endochironomus impar-type	–	3	–	–	3	8	–
Endochironomus tendens-type	–	–	–	1	–	–	–
Glyptotendipes pallens-type	–	–	–	2	–	–	–
Harnischia	–	–	–	3	–	–	–
Lauterborniella	–	–	–	–	1	–	–
Microtendipes pedellus-type	1	4	1	3	3	1	–
Omisus	–	1	–	–	–	–	–
Pagastiella	–	4	–	–	1	2	–
Parachironomus varus-type	–	–	–	1	–	–	–
Paracladopelma	–	–	–	1	–	–	–

(*cont.*)

Country		SK			PL	UA	RO
Lake district	**Lt**	**Ta**	**Vi**	**St**	**Ta**	**Uc**	**Ma**
Paratendipes albimanus-type	1	–	1	1	–	–	–
Phaenopsectra flavipes-type	–	1	1	–	–	–	–
Polypedilum nubeculosum-type	–	2	1	3	1	1	–
Polypedilum sordens-type	–	2	–	2	–	–	–
Sergentia coracina-type	1	1	1	1	–	–	–
Stictochironomus type B	–	–	–	–	1	–	–
Chironominae – Tanytarsini							
Cladotanytarsus mancus-type	–	–	1	3	–	4	–
Micropsectra contracta-type	1	2	–	2	2	–	–
Micropsectra insignilobus-type	1	5	–	–	1	–	–
Micropsectra junci-type	1	1	–	–	1	2	–
Micropsectra pallidula-type	–	–	–	1	–	–	–
Micropsectra radialis-type	–	23	–	–	10	1	–
Micropsectra type A	1	6	–	–	4	3	–
Paratanytarsus austriacus-type	1	21	–	–	1	1	–
Paratanytarsus pencillatus-type	–	3	–	3	4	11	–
Stempellina	–	–	–	3	–	–	–
Tanytarsus lugens-type	1	23	1	–	8	6	1
Tanytarsus mendax-type	1	11	–	3	4	8	–
Tanytarsus nemorosus-type	–	–	1	–	–	–	–
Tanytarsus pallidicornis-type	1	1	–	–	2	–	–
Tanytarsini indet.	–	–	–	–	1	–	–

Appendix 2

List of genera present in the book along with the genera that's presence in lake sediments in the Carpathians is likely.

		Western Carpathians			Ukrainian Carpathians	Southern Carpathians	
	Present in the book	Novikmec et al. (2015)	Bitušík et al. (2006)	Hamerlík et al. (2019)	Bitušík et al. (2020)	Tatole (2023)	Diaconu et al. (2017)
Podonominae							
Parochlus	+						
Tanypodinae							
Ablabesmyia	+						
Apsectrotanypus			+				
*Conchapelopia**						+	
Derotanypus	+						
*Krenopelopia**						+	
Macropelopia	+						
Monopelopia							+
*Paramerina***						+	
Procladius	+						
Psectrotanypus						+	
Zavrelimyia	+						
Podonominae							
Lasiodiamesa				+			
Diamesinae							
*Boreoheptagyia**						+	
Diamesa	+						
Pseudodiamesa	+						
Pseudokiefferiella	+						
Prodiamesinae							
Prodiamesa	+						

 | DOI:10.1163/9789004694781_011

(*cont.*)

		Western Carpathians			Ukrainian Carpathians	Southern Carpathians	
	Present in the book	Novikmec et al. (2015)	Bitušík et al. (2006)	Hamerlík et al. (2019)	Bitušík et al. (2020)	Tatole (2023)	Diaconu et al. (2017)
Orthocladiinae							
Acamptocladius					+		
Brillia	+						
Chaetocladius	+						
Corynoneura	+						
Cricotopus	+						
Diplocladius	+						
Epoicocladius	+						
Eukiefferiella	+						
Georthocladius						+	+
Gymnometriocnemus / Bryophaenocladius	+						
*Heleniella**			+			+	
Heterotrissocladius	+						
Hydrobaenus	+						
*Krenosmittia**			+			+	
Limnophyes / Paralimnophyes	+						
Metriocnemus	+						
Nanocladius	+						
Orthocladius	+						
Parachaetocladius		+					
Paracladius	+						
Paracricotopus						+	
Parakiefferiella	+						
Parametriocnemus / Paraphaenocladius	+						
Paratrissocladius	+						

(*cont.*)

		Western Carpathians			Ukrainian Carpathians	Southern Carpathians	
	Present in the book	**Novikmec et al. (2015)**	**Bitušík et al. (2006)**	**Hamerlík et al. (2019)**	**Bitušík et al. (2020)**	**Tatole (2023)**	**Diaconu et al. (2017)**
Parorthocladius	+						
Psectrocladius	+						
Pseudorthocladius							+
Pseudosmittia	+						
Rheocricotopus	+						
Rheosmittia						+	
Smittia / Parasmittia	+						
Stilocladius		+					
Synorthocladius	+						
Thienemanniella	+						
*Tokunagaia*****		+					
Tvetenia	+						
Zalutschia	+						
Chironominae – Chironomini							
Benthalia	+						
Chironomus	+						
Cladopelma	+						
Cryptochironomus	+						
Cryptotendipes	+						
Dicrotendipes	+						
Endochironomus / Synendotendipes	+						
Glyptotendipes	+						
Harnischia	+						
Lauterborniella	+						
Microtendipes	+						
Omisus	+						
Pagastiella	+						

(*cont.*)

		Western Carpathians			Ukrainian Carpathians	Southern Carpathians	
	Present in the book	**Novikmec et al. (2015)**	**Bitušík et al. (2006)**	**Hamerlík et al. (2019)**	**Bitušík et al. (2020)**	**Tatole (2023)**	**Diaconu et al. (2017)**
Parachironomus	+						
Paracladopelma	+						
Paratendipes	+						
Phaenopsectra	+						
Polypedilum	+						
Sergentia	+						
Stictochironomus	+						
*Synendotendipes****		+					
Zavrelia						+	
Chironominae – Tanytarsini	+						
Cladotanytarsus	+						
Micropsectra	+						
Parapsectra *****						+	
Paratanytarsus	+						
Rheotanytarsus						+	
Stempellina	+						
Stempellinella						+	
Tanytarsus	+						

* Rheobiont
** Currently moved to *Zaverlimyia* as subgenus
*** In this book listed among *Endochironomus* as *E. impar*-type
**** In this book as *Eukiefferiella clarpennis*-type T
***** Currently moved to *Micropsectra*

References

Alm, T. & Willassen, E. (1993) Late Weichselian Chironomidae (Diptera) stratigraphy of Lake Nedre Æråsvatn, Andøya, Northern Norway. *Hydrobiologia* 264(1): 21–32.

Andersen, T., Baranov, V., Hagenlund, L.K., Ivković, M., Kvifte, G.M. & Pavlek, M. (2016) Blind Flight? A New Troglobiotic Orthoclad (Diptera, Chironomidae) from the Lukina Jama – Trojama Cave in Croatia. *PLoS ONE* 11(4): e0152884.

Andersen, T., Cranston, P.S. & Epler, J.H. (2013a) The larvae of Chironomidae (Diptera) of the Holarctic Region. Keys and diagnoses. *Insect Systematics & Evolution* 66: 1–571.

Andersen, T., Sæther, O.A., Cranston, P.S. & Epler, J.H. (2013b) 9. The larvae of Orthocladiinae (Diptera: Chironomidae) of the Holarctic Region – Keys and diagnoses. In Andersen, T., Sæther, O.A., Cranston, P.S. & Epler, J.H. (eds), Chironomidae of the Holarctic Region. Keys and diagnoses. Larvae. *Insect Systematics & Evolution*, Suppl. 66: 189–386.

Anderson, N.J., Brodersen, K.P., Ryves, D.B., McGowan, S., Johansson, L.S., Jeppesen, E. & Leng, M.J. (2008) Climate versus in-lake processes as controls on the development of community structure in a low-arctic lake (South-West Greenland). *Ecosystems* 11(2): 307–324.

Armitage, P.D. (1995) Behaviour and ecology of adults. In Armitage, P.D., Cranston, P.S., Pinder, L.C.V. (eds), *The Chironomidae: Biology and ecology of non-biting midges*. Chapman Hall, London, Glasgow, Weinheim N.Y., Melbourne, Madras, p. 195–224.

Armitage, P.D., Pinder, L.C. & Cranston, P.S. (eds), (1995) *The Chironomidae: biology and ecology of non-biting midges*. Chapman Hall, London, Glasgow, Weinheim N.Y., Melbourne, Madras.

Bálint, M., Ujvárosi, L., Theissinger, K., Lehrian, S., Mészáros, N. & Pauls, S.U. (2011) The Carpathians as a Major Diversity Hotspot in Europe. In Zachos, F.E. & Habel, J.C. (eds), *Biodiversity Hotspots*, Springer-Verlag Berlin, Heidelberg. p. 189–205.

Balteanu, D., Nastase, M., Dumitrascu, M. & Grigorescu, I. (2016) Environmental changes in the Maramures Mountains Natural Park. In Zhelezov, G. (ed.), *Sustainable Development in Mountain Regions*, Springer, Cham Heidelberg New York Dordrecht London, p. 335–348.

Baranov, V. (2011) A preliminary annotated checklist of non-biting midges (Diptera, Chironomidae) of Ukraine. *Ukrainska Entomofaunistyka* 2(1): 7–24. [In Ukrainian].

Bauernfeind, E. & Humpesch, U.H. (2001) *Die Eintagsfliegen Zentraleuropas (Insecta: Ephemeroptera): Bestimmung und Ökologie*. Naturhistorisches Museum, Wien, 239 pp.

Bazzanti, M., Seminara, M., Baldoni, S. & Dowgiallo, M.G. (1998) Assessing hypolimnetic stress in a monomictic, eutrophic lake using profundal sediment and macrobenthic characteristics. *Journal of Freshwater Ecology* 13(4): 405–412.

Berg, M.B. (1995) Larval food and feeding behaviour. In: Armitage, P.D., Pinder, L.C. & Cranston, P.S. (eds), *The Chironomidae: biology and ecology of non-biting midges*. Chapman & Hall, 572 pp.

Birks, H.H. & Birks, H.J.B. (2006) Multi-proxy studies in palaeolimnology. *Veg. His. Archaeobot.* 15(4): 235–251.

Bitušík, P. (2000) *Príručka na určovanie lariev pakomárov (Diptera. Chironomidae) Slovenska*. Časť 1. *Buchonomyinae, Diamesinae, Prodiamesinae a Orthocladiinae*. Vydavateľstvo Technickej Univerzity vo Zvolene, 133 pp. (in Slovak).

Bitušík, P. (2004) Chironomids (Diptera. Chironomidae) of the mountain lakes in the Tatra Mts. (Slovakia). A review. *Dipterologica bohemoslovaca 12*. *Acta Facultatis Ecologiae (Zvolen)* 12, Suppl. 1: 25–53.

Bitušík, P. & Hamerlík, L. (2014) *Príručka na určovanie lariev pakomárov (Diptera: Chironomidae) Slovenska*. Časť 2. *Tanypodinae*. Vydavateľstvo Univerzity Matej Bela v Banskej Bystrici – Belianum, 96 pp.

Bitušík, P. & Kubovčík, V. (1999) Sub-fossil chironomids (Diptera: Chironomidae) from the sediments of the Nižné Terianske pleso (High Tatra Mts., Slovakia). *Dipterologica Bohemoslovaca* 9: 11–20.

Bitušík, P., Dobríková, D., Pipík, R. & Hamerlík, L. (2017) Relict chironomid communities surviving in the coldest High Tatra Mountain lakes confirmed by a palaeolimnological survey. *Biologia* 72(8): 965–969.

Bitušík, P., Kubovčík, V., Štefková, E., Appleby, P.G. & Svitok, M. (2009) Subfossil diatoms and chironomids along an altitudinal gradient in the High Tatra Mountain lakes: a multi-proxy record of past environmental trends. *Hydrobiologia* 631: 65–85.

Bitušík, P., Novikmec, M., & Hamerlik, L. (2020) Chironomids (Insecta, Diptera, Chironomidae) from alpine lakes in the Eastern Carpathians with comments on newly-recorded species from Ukraine. *Biodiversity Data Journal* 8.

Bitušík, P., Šporka, F. & Krno, I. (2010) Benthic macroinvertebrate fauna of two alpine lakes over the last century: The value of historical data for interpreting environmental changes. *Biologia* 65(5): 884–891.

Bitušík, P., Svitok, M. & Bačík, J. (2007) Chironomids (Diptera: Chironomidae) of man-made reservoirs in Banská Štiavnica mining region (Slovakia). *Acta Zoologica Universitatis Comenianae* 47: 15–26.

Bitušík, P., Svitok, M., Kološta, P. & Hubková, M. (2006a) Classification of the Tatra Mountains lakes (Slovakia) using chironomids (Diptera, Chironomidae). *Biologia* 61, Suppl. 18: 191–201.

Bitušík, P., Kopáček, J., Stuchlík, E. & Šporka, F. (eds) (2006b) Limnology of Lakes in the Tatra Mountains. *Biologia* 61, Suppl. 18: 1–222.

Bitušík, P., Trnková, K., Chamutiová, T., Sochuliaková, L., Stoklasa, J., Pipík, R., Szarlowicz, K., Szacilowski, G., Thomková, K., Šporka, F., Starek, D., Šurka, J., Milovský, R. & Hamerlík, L. (2018) Tracking human impact in a mining landscape using lake sediments: a multi-proxy palaeolimnological study. *Palaeogeography, Palaeoclimatology, Palaeoecology* 504: 23–33.

Boggero, A., Füreder, L., Lencioni, V., Simcic, T., Thaler, B., Ferrarese, U., Lotter, A.F. & Ettinger, R. (2006) Littoral chironomid communities of Alpine lakes in relation to environmental factors. *Hydrobiologia* 562: 145–165.

Botnariuc, N. & Tatole, V. (1997) Considerations on the chironomid fauna from the Retezat Massif. *Ocrot. Nat. med. înconj.* 41: 23–33.

Brodersen, K.P. & Anderson, N.J. (2002) Distribution of chironomids (Diptera) in low arctic West Greenland lakes: trophic conditions, temperature and environmental reconstruction. *Freshwater Biology* 47(6): 1137–1157.

Brodersen, K.P. & Quinlan, R. (2006) Midges as palaeoindicators of lake productivity, eutrophication and hypolimnetic oxygen. *Quaternary Science Reviews* 25(15–16): 1995–2012.

Brodersen, K.P., Dall, P.C. & Lindegaard, C. (1998) The fauna in the upper stony littoral of Danish lakes: macroinvertebrates as trophic indicators. *Freshwater Biology* 39(3): 577–592.

Brodersen, K.P., Lindegaard, C. & Anderson, N.J. (2001a) Holocene temperature and environmental reconstruction from lake sediments in the Søndre Strømfjord region, southern West Greenland. *Geological Survey of Denmark and Greenland Bulletin* p. 59–64.

Brodersen, K.P., Odgaard, B.V., Vestergaard, O. & Anderson, N.J. (2001b) Chironomid stratigraphy in the shallow and eutrophic Lake Søbygaard, Denmark: chironomid – macrophyte co-occurrence. *Freshwater Biology* 46(2): 253–267.

Brodersen, K.P., Pedersen, O., Lindegaard, C. & Hamburger, K. (2004) Chironomids (Diptera) and oxy-regulatory capacity: An experimental approach to paleolimnological interpretation. *Limnology and Oceanography* 49: 1549–1559.

Brodin, Y.W. (1986) The postglacial history of Lake Flarken, Southern Sweden, interpreted from subfossil insect remains. *Internationale Revue der Gesamten Hydrobiologie* 71: 371–432.

Brodin, Y.W. & Gransberg, M. (1993) Responses of insects, especially Chironomidae (Diptera), and mites to 130 years of acidification in a Scottish lake. *Hydrobiologia* 250(3): 201–212.

Brooks, S.J., Bennion, H. & Birks, H.J.B. (2001) Tracing lake trophic history with a chironomid – total phosphorus inference model. *Freshwater Biology* 46(4): 513–533.

Brooks, S.J. & Birks, H.J.B. (2001) Chironomid-inferred air temperatures from Lateglacial and Holocene sites in north-west Europe: progress and problems. *Quaternary Science Reviews* 20(16–17): 1723–1741.

Brooks, S.J. & Birks, H.J.B. (2004) The dynamics of Chironomidae (Insecta: Diptera) assemblages in response to environmental change during the past 700 years on Svalbard. *Journal of Paleolimnology* 31(4): 483–498.

Brooks, S.J., Langdon, P.G. & Heiri, O. (2007) *The Identification and Use of Palaearctic Chironomidae Larvae in Palaeoecology*. QRA Technical Guide No. 10. QRA, London, 276 pp.

Brundin, L. (1956) Die bodenfaunistischen Seentypen und ihre Anwendbarkeit auf die Südhalbkugel. Zugleich eine Theorie der produktionsbiologischen Bedeutung der glazialen Erosion. *Rep. Inst. Freshwater. Res.* 37: 186–235.

Brundin, L. (1966) Transantarctic relationships and their significance, as evidenced by chironomid midges. With a monograph on the subfamilies Podonominae and Aphroteniinae and the austral Heptagyiae. *Kunglica Svenska Vetenskapsakademiens Handlingar* 11: 1–472.

Brundin, L. (1983) The larvae of Podonominae (Diptera: Chironomidae) of the Holarctic region – Keys and diagnoses. *Entomologica Scandinavica*, Suppl. 19: 23–31.

Buczkó, K., Korponai, J., Máthé, I., Jakab, G., Tóth, M., Braun, M., Ács, É., Karátson, D., Magyari, E.K. (2022) Limnological Changes in Lake Sf. Ana. In Ciomadul (Csomád), The Youngest Volcano in the Carpathians: Volcanism, Palaeoenvironment, Human Impact. Cham: Springer International Publishing, p. 159–174.

Buczkó, K., Magyari, E.K., Bitušík, P. & Wacnik, A. (2009) Review of dated Late Quaternary palaeolimnological records in the Carpathian Region, east-central Europe. *Hydrobiologia* 631(1): 3–28.

Cankov, C. (1974) The Bulgarian Carpathian-Balkan region. The Southern Carpathians. In: Maheľ, M. (ed.), *Tectonics of the Carpathian-Balkan regions*, Veda, Bratislava, p. 303–307.

Cereghino, R., Biggs, J., Oertli, B., Declerck, S. (2008) The ecology of European ponds: defining the characteristics of a neglected freshwater habitat. *Hydrobiologia* 597: 1–6.

Chamutiová, T., Hamerlik, L. & Bitušík, P. (2020) Subfossil chironomids (Diptera, Chironomidae) of lakes in the Tatra Mountains: an illustrated guide. *Zootaxa* 4819(2): 216–264.

Chamutiová, T., Hamerlík, L., Szarlowicz, K., Pipik, R., Starek, D., Šporka, F., Přidalová, M., Sochuliaková, L. & Bitušík, P. (2018) The historical development of three man-made reservoirs in a mining region: a story told by subfossil chironomids. *Journal of Limnology* 77(s1): 220–229.

Chernyshev, I.V., Konečný, V., Lexa, J., Kovalenker, V.A., Jeleň, S., Lebedev, V. & Goltsman, Y.V. (2013) K-Ar and Rb-Sr geochronology and evolution of the Štiavnica Stratovolcano (Central Slovakia). *Geologica Carpathica* 64: 327–351.

Cogălniceanu, D., Tudorancea, M., Preda, E., & Gâldean, N. (2009) Evaluating diversity of chironomid (Insecta: Diptera) communities in alpine lakes, Retezat National Park (Romania). *Advances in Limnology* 62: 191–213.

Contreras-Lichtenberg, R. (1984) Revision der in der Westpaläarktis verbreiteten Arten des Genus Dicrotendipes Kieffer, 1913 (Diptera, Nematocera, Chironomidae).

Annalen des Naturhistorischen Museums in Wien. Serie B für Botanik und Zoologie 1: 663–726.

Courtney-Mustaphi, C.J., Steiner, E., von Fumetti, S., Heiri, O. (2024) Aquatic invertebrate mandibles and sclerotized remains in Quaternary lake sediments. *Journal of Paleolimnology* 71: 45–83.

Cranston, P.S. 1995. Introduction. In: Armitage, P.D., Cranston, P.S., Pinder, L.C.V. (eds), The Chironomidae. Biology and ecology of non-biting midges. Chapman Hall, Lond., Glasgow, Weinheim, N.Y., Melbourne, Madras, p. 1–7.

Cranston, P.S. & Epler, J.H. (2013) 5. The larvae of Tanypodinae (Diptera: Chironomidae) of the Holarctic region – Keys and diagnoses. In: Andersen, T., Sæther, O.A., Cranston, P.S. & Epler, J.H. (eds), Chironomidae of the Holarctic Region. Keys and diagnoses. Larvae. *Insects systematics & evolution*, Suppl. 66: 39–136.

Cranston, P.S. & Krosch, M.N. (2015) DNA sequences and austral taxa indicate generic synonymy of Paratrichocladius Santos-Abreu with *Cricotopus* Wulp (Diptera: Chironomidae). *Systematic Entomology* 40(4): 719–732.

Cuppen, H.P.J.J. & Tempelman, D. (2018) Identification key for the 4th stage larvae of north west European species of Cricotopus (Diptera: Chironomidae: Orthocladiinae). *Lauterbornia* 85: 69–90.

Currie, D.C. & Walker, I.R. (1992) Recognition and palaeohydrologic significance of fossil black fly larvae, with a key to the Nearctic genera (Diptera: Simuliidae). *Journal of Paleolimnology* 7(1): 37–54.

DeWalt, R.E., Kondratieff, B.C. & Sandberg, J.B. (2015) Order Plecoptera. In Ecology and General Biology: Thorp and Covich's Freshwater Invertebrates; Thorp, J. & Rogers, D.C. (eds), Academic Press: Cambridge, MA, USA, p. 933–949.

DeWalt, R.E. & Ower, G.D. (2019) Ecosystem services, global diversity, and rate of stonefly species descriptions (Insecta: Plecoptera). *Insects* 10(4): 99.

Diaconu, A.C., Tóth, M., Lamentowicz, M., Heiri, O., Kuske, E., Tanţau, I., Panait, A.M., Braun, M., Feurdean, A. (2017) How warm? How wet? Hydroclimate reconstruction of the past 7500 years in northern Carpathians, Romania. *Palaeogeography, Palaeoclimatology, Palaeoecology* 482: 1–12.

Dieffenbacher-Krall, A.C., Vandergoes, M.J., Woodward, C.A. & Boothroyd, I.K.G. (2008) *Guide to Identification and Ecology of New Zealand Subfossil Chironomids Found in Lake Sediment.* Climate Change Institute, University of Maine, Orono, ME (http://www.climatechange.umaine.edu/Research/facilities/perl/nzguide.html).

Dimitriadis, S. & Cranston, P.S. (2001) An Australian Holocene climate reconstruction using Chironomidae from a tropical volcanic maar lake. *Palaeogeography, Palaeoclimatology, Palaeoecology* 176(1–4): 109–131.

Eggermont, H. & Heiri, O. (2012) The chironomid-temperature relationship: expression in nature and palaeoenvironmental implications. *Biological Reviews* 87(2): 430–456.

Eggermont, H. & Verschuren, D. (2004a) Sub-fossil Chironomidae from East Africa. 1. Tanypodinae and Orthocladiinae. *Journal of Paleolimnology* 32: 383–412.

Eggermont, H. & Verschuren, D. (2004b) Sub-fossil Chironomidae from East Africa. 2. Chironominae (Chironomini and Tanytarsini). *Journal of Paleolimnology* 32: 413–455.

Ekrem, T. (2004) Immature stages of Europaean Tanytarsus species I. The eminulus-, gregarius-, lugens- and mendax species groups (Diptera, Chironomidae). *Deutsche Entomologische Zeitschrift* 51: 97–146.

Elias, S.A. (2010) Coleoptera and Trichoptera. In: Smol, J.P., Birks, H.J.B. & Last, W.M. (eds). *Tracking environmental change using late sediments*. Vol. 4. Zoological Indicators, Kluwer Academic Publishers, Dordrecht, p. 67–80.

Engels, S., Bohncke, S.J.P., Bos, J.A.A., Brooks, S.J., Heiri, O. & Helmens, K.F. (2008) Chironomid-based palaeotemperature estimates for northeast Finland during Oxygen Isotope Stage 3. *Journal of Paleolimnology* 40(1): 49–61.

Epler, J.H. (1995) *Identification manual for the larval Chironomidae (Diptera) of Florida*. (Revised Edition).

Epler, J.H., Ekrem, T. & Cranston, P.S. (2013) 10. The larvae of Chironominae (Diptera: Chironomidae) of the Holarctic Region – Keys and diagnoses. In: Andersen, T., Sæther, O.A., Cranston, P.S. & Epler, J.H. (eds), Chironomidae of the Holarctic Region. Keys and diagnoses. Larvae. *Insect Systematics & Evolution*, Suppl. 66: 387–556.

Fărcaş, S., Tanţău, I., Mîndrescu, M. & Hurdu, B. (2013) Holocene vegetation history in the Maramureş mountains (Northern Romanian Carpathians). *Quaternary International* 293: 92–104.

Ferrington, L.C. (2007) Global diversity of non-biting midges (Chironomidae; Insecta-Diptera) in freshwater. *Hydrobiologia* 595: 447–455.

Ferrington, L.C. & Sæther, O.A. (2011) A revision of the genera *Pseudosmittia* Edwards, 1932, *Allocladius* Kieffer, 1913, and *Hydrosmittia* gen. n. (Diptera: Chironomidae, Orthocladiinae). *Zootaxa* 2849(1): 1–314.

Filippi, M.L., Heiri, O., Arpenti, E., Angeli, N., Bortolotti, M., Lotter, A.F. & Van Der Borg, K. (2007) Studio paleolimnologico del Lago Nero di Cornisello (Parco Naturale Adamello-Brenta, Trentino). *Studi Trent. Sci. Nat., Acta Geol.* 82(2005): 261–278.

Finnie, T.J.R., Preston, C.D., Hill, M.O., Uotila, P., Crawley, M.J. 2007. Floristic elements in European vascular plants: an analysis based on Atlas Florae Europaeae. *Journal of Biogeography* 34: 1848–1872.

Fittkau, E.J. & Roback, S.S. (1983) The larvae of Tanypodinae (Diptera: Chironomidae) of the Holarctic region – Keys and diagnoses. In: Wiederholm, T. (ed.), Chironomidae of the Holarctic region. Keys and diagnoses. Part 1. Larvae. *Entomologica Scandinavica*, Suppl. 19: 33–110.

Füreder, L., Ettinger, R., Boggero, A., Thaler, B. & Thies, H. (2006) Macroinvertebrate diversity in Alpine lakes: effects of altitude and catchment properties. *Hydrobiologia* 562: 123–144.

Gandouin, E. & Franquet, E. (2002) Late Glacial and Holocene chironomid assemblages in Lac Long Inférieur (southern France, 2090 m): palaeoenvironmental and palaeoclimatic implications. *Journal of Paleolimnology* 28(3): 317–328.

Gąsiorowski, M. & Sienkiewicz, E. (2010) The Little Ice Age recorded in sediments of a small dystrophic mountain lake in southern Poland. *Journal of Paleolimnology* 43: 475–487.

Găştescu, P. (2010) The lakes in Romania – An actual synthesis. *Lakes, reservoirs and ponds* 4(1): 15–23.

Giłka, W. (2005) A systematic review of European *Stempellina* Thienemann et Bause, 1913 (Diptera: Chironomidae) with description of a new species from Fennoscandia. *Annales Zoologici* 55(3): 413–419.

Giłka, W. (2007) Przeglad faunistyczny ochotkowatych z plemienia Tanytarsini (Diptera: Chironomidae) Tatrzanskiego Parku Narodowego. A faunistic review of chironomids of the tribe Tanytarsini (Diptera: Chironomidae) of the Tatra National Park. *Dipteron* 23: 11–17.

Gorek, A. & Kahan, Š. (1973) Prehľad geologického vývoja a stavby Vysokých Tatier. *Zborník* TANAP 15: 5–88.

Gurung, A.B., Bokwa, A., Chełmicki, W., Elbakidze, M., Hirschmugl, M., Hostert, P., Ibisch, P., Kozak, J., Kuemmerle, T., Matei, E., Ostapowicz, K., Pociask-Karteczka, J., Schmidt, L., van der Linden, S.& Zebisch, M. 2009. Global change research in the Carpathian mountain region. *Mountain Research and Development* 29(3), 282–289.

Hájková, P., Pařil, P., Petr, L., Chattová, B., Grygar, T.M. and Heiri, O. (2016) A first chironomid-based summer temperature reconstruction (13-5 ka BP) around 49 N in inland Europe compared with local lake development. *Quaternary Science Reviews* 141: 94–111.

Hamerlík, L. & Bitušík, P. (2008) Records of Chironomidae (Diptera) from a subalpine Tatra lake, new in Slovakia. *Lauterbornia* 62: 131–133.

Hamerlík, L. & Bitušík, P. (2009) The distribution of littoral chironomids along an altitudinal gradient in High Tatra Mountain lakes: Could they be used as indicators of climate change? *Annales de Limnologie-International Journal of Limnology* 45: 145–156.

Hamerlík, L. & Brodersen, K.P. (2010) Non-biting midges (Diptera: Chironomidae) from fountains of two European cities: micro-scale island biogeography. *Aquatic Insects* 32(1): 67–79.

Hamerlík, L. & Silva, F.L. (2021) *Chironomidae of Central America. An illustrated Introduction of Larval Subfossils*. CRC Press, Taylor & Francis Group, 186 pp.

Hamerlík, L., Dobríková, D., Szarlowicz, K., Reczynski, W., Kubica, B., Šporka, F. & Bitušík, P. (2016) Lake biota response to human impact and local climate during the last 200 years: A multi-proxy study of a subalpine lake (Tatra Mountains, W Carpathians). *Science of the Total Environment* 545–546: 320–328.

Hamerlík, L., Silva, F.L. & Massaferro, J. (2022) An illustrated guide of subfossil Chironomidae (Insecta: Diptera) from waterbodies of Central America and the Yucatan Peninsula. *Journal of Paleolimnology* 67(3): 201–258.

Hamerlík, L., Svitok, M., Novikmec, M., Očadlík, M., Bitušík, P. (2014) Local, among-site, and regional diversity patterns of benthic macroinvertebrates in high altitude waterbodies: do ponds differ from lakes? *Hydrobiologia* 723: 41–52.

Hamerlík, L., Svitok, M., Novikmec, M., Veselská, M. & Bitušík, P. (2017) Weak altitudinal pattern of overall chironomid richness is a result of contrasting trends of subfamilies in high-altitude ponds. *Hydrobiologia* 793(1): 67–81.

Heiri, O. & Lotter, A.F. (2003) 9000 years of chironomid assemblage dynamics in an Alpine lake: long-term trends, sensitivity to disturbance, and resilience of the fauna. *Journal of Paleolimnology* 30(3): 273–289.

Heiri, O. & Lotter, A.F. (2007) Sciaridae in lake sediments: indicators of catchment and stream contribution to fossil insect assemblages. *Journal of Paleolimnology* 38(2): 183–189.

Heiri, O., Lotter, A.F. (2010) How does taxonomic resolution affect chironomid-based temperature reconstruction? *Journal of Paleolimnology* 44: 589–601.

Heiri, O., Wick, L., van Leeuwen, J.F., van der Knaap, W.O. & Lotter, A.F. (2003) Holocene tree immigration and the chironomid fauna of a small Swiss subalpine lake (Hinterburgsee, 1515 m asl). *Palaeogeography, Palaeoclimatology, Palaeoecology* 189(1–2): 35–53.

Henrikson, L., Olofsson, J.B. & Oscarson, H.G. (1982) The impact of acidification on Chironomidae (Diptera) as indicated by subfossil stratification. *Hydrobiologia* 86(3): 223–229.

Herenchuk, K.I. (ed.) (1968) Nature of the Ukrainian Carpathians. Vydavnytstvo Lvivskoho Universytetu, Lviv, Ukraine (in Ukrainian).

Hofmann, W. (1971) Zur Taxonomie und Palökologie subfossiler Chironomiden (Dipt.) in Seesedimenten. *Archiv für Hydrobiologie Beihefte Ergebnisse der Limnologie* 6: 1–50.

Holzenthal, R.W., Morse, C.J., Kjer, K.M. (2011) Order Trichoptera Kirby, 1813. In Zhang, Z.Q. (ed.), Animal Biodiversity: An outline of Higher-level Classification and Survey of Taxonomic Richness. *Zootaxa* 3148: 209–211.

Holubets, M.A., Honchar, M.T., Komendar, V.I., Kutsheraviy, V.A. & Odynak, J.P. (eds) (1988) *Nature of the Ukrainian Carpathians*. Naykova Dumka, Kyiv, Ukraine (in Ukrainian).

Hrabě, S. 1939. *Bentická zvířena tatranských jezer*. Sborník Klubu Přírodověd. v Brně 22: 1–13.

Hrabě, S. 1942. O bentické zvířeně jezer ve Vysokých Tatrách. *Physiographica Slovaca, Acta Eruditae Societatis Slovacae* 8: 124–177.

Ilkova, J., Michailova, P., Szarek-Gwiazda, E., Kownacki, A. & Ciszewski, D. (2017) The response of genome of the chironomidae (Diptera) to heavy metal pollution in two rivers of Southern Poland. *Acta Zoologica Bulgarica* 8: 9–15.

Ilyashuk, E.A., Ilyashuk, B.P., Hammarlund, D., Larocque, I. (2005) Holocene climatic and environmental changes inferred from midge records (Diptera: Chironomidae, Chaoboridae, Ceratopogonidae) at Lake Berkut, southern Kola Peninsula, Russia. *Holocene* 15: 897–914.

Ilyashuk, B.P., Ilyashuk, E.A., Makarchenko, E.A. & Heiri, O. (2010) Midges of the genus *Pseudodiamesa* Goetghebuer (Diptera, Chironomidae): current knowledge and palaeoecological perspective. *Journal of Paleolimnology* 44(2): 667–676.

Ilyashuk, E.A., Koinig, K.A., Heiri, O., Ilyashuk, B.P. & Psenner, R. (2011) Holocene temperature variations at a high-altitude site in the Eastern Alps: a chironomid record from Schwarzsee ob Sölden, Austria. *Quaternary Science Reviews* 30(1–2): 176–191.

Jambrović, M. (2022) *Subfossil Chironomidae as paleoindicators of past environmental changes: a case study of a glacial lake in the Low Tatra Mountains*. Master's thesis, 50 pp.

Janecek, B. & Moog, O. (2002) Diptera: Chironomidae. In: Moog, O. (ed.). *Fauna Aquatica Austriaca*, Edition 2002, *Wassserwirtschaftskataster, Bundesministerium für Land und Forstwirtschaft. Umwelt und Wasserwirtschaft*, Vienna.

Johnson, R.K. (1989) Classification of profundal chironomid communities in oligotrophic/humic lakes of Sweden using environmental data. *Acta Biol. Debr. Oecol. Hung*. 3: 167–175.

Kadlečík, J. (ed.) 2016. *Carpathians – Hidden Treasures*. Štátna ochrana prírody, Banská Bystrica, 164 pp.

Kłapyta, P., Zasadni, J., Pociask-Karteczka, J., Gajda, A., & Franczak, P. (2016) Late Glacial and Holocene paleoenvironmental records in the Tatra Mountains, East-Central Europe, based on lake, peat bog and colluvial sedimentary data: A summary review. *Quaternary International* 415: 126–144.

Kobiv, Y., Prokopiv, A., Borsukevich, L., Helesh, M. (2009) Distribution, state of populations and habitat characteristics of rare and endangered species of plants in the northern part of the Svydovets Mountains (Ukrainian Carpathians). *Visnyk of the Lviv University. Series Biology* 49: 83–90.

Kondracki, J. (1989) *Karpaty*. Warszawa, Wydawnictwa Szkolne i Pedagogiczne, 262 pp.

Kondracki, J.A. *Carpathian Mountains*. Encyclopedia Britannica, 4 Apr. 2022, https://www.britannica.com/place/Carpathian-Mountains. Accessed 29 July 2022.

Konečný, V., Lexa, J., Halouzka, R., Hók, J., Vozár, J., Dublan, L., Nagy, A., Šimon, L., Havrila, M., Ivanička, J., Hojstričová, V., Mihaliková, A., Vozárová, A., Konečný, P., Kováčiková, M., Filo, M., Marcin, D., Klukanová, A., Liščák, P. & Žáková, E. (1998) *Explanatory notes to the geological map of Štiavnické vrchy and Pohronský Inovec mountain ranges (Štiavnica Stratovolcano)*. Geological Survey of Slovak Republic, Bratislava, 473 pp. (in Slovak with English summary).

Koščo, J., Košuthová, L., Košuth, P. & Pekárik, L. (2010) Non-native fish species in Slovak waters: origins and present status. *Biologia* 65(6): 1057–1063.

Kowalyk, H.E. (1985) The larval cephalic setae in the Tanypodinae (Diptera: Chironomidae) and their importance in generic determinations. *The Canadian Entomologist* 117: 67–106.

Kownacki, A. (1987) Benthic invertebrate fauna of high mountain streams in Caucasus. *Acta Universitastis Lodziensis, Folia Caucasica* 2: 89–99.

Kownacki, A. (2010) Chironomidae (Diptera, Insecta) of the Tatra National Park: distribution, ecology, zoogeography. (Chironomidae (Diptera, Insecta) Tatrzańskiego Parku Narodowego rozmieszczenie, ekologia, zoogeografia). *Nauka a zarządzanie obszarem Tatr i ich otoczeniem* 2: 113–118. (In Polish).

Kownacki, A. (2011) Significance and conservation of Chironomidae (Diptera, Insecta) in aquatic ecosystems of Poland (Znaczenie i ochrona Chironomidae (Diptera, Insecta) w ekosystemach wodnych Polski). *Forum Faunistyczne* 1(1): 4–11. (In Polish with English Summary).

Král, V. 2001. *Fyzická geografie Evropy*. Academia, Praha, 350 pp.

Križáni, I., Andráš, P. & Ladomerský, J. (2007) *Contamination of Štiavnicé vrchy Mountains by mining activity*. Technical University, Zvolen, 100 pp. (in Slovak with English summary).

Krno, I.J., Šporka, F., Galas, J., Hamerlík, L., Zaťovičová, Z. & Bitušík, P. (2006) Littoral benthic macroinvertebrates of mountain lakes in the Tatra Mountains (Slovakia, Poland). *Biologia* 61(18): S147–S166.

Kubovčík, V. (2012) *Palaeoecology of Tatra Mts. lakes: Chironomids (Diptera: Chironomidae), climate change and acidification*. Technical University in Zvolen, 95 pp. (in Slovak).

Kubovčík, V. & Bitušík, P. (2006) Subfossil chironomids (Diptera, Chironomidae) in three Tatra Mountain lakes (Slovakia) on an acidification gradient. *Biologia* 61(18): S213–S220.

Langdon, P.G., Ruiz, Z.O.E., Brodersen, K.P. & Foster, I.D. (2006) Assessing lake eutrophication using chironomids: understanding the nature of community response in different lake types. *Freshwater Biology* 51(3): 562–577.

Langdon, P.G., Ruiz, Z.O.E., Wynne, S., Sayer, C.D. & Davidson, T.A. (2010) Ecological influences on larval chironomid communities in shallow lakes: implications for palaeolimnological interpretations. *Freshwater Biology* 55(3): 531–545.

Larocque, I. & Hall, R.I. (2003) Chironomids as quantitative indicators of mean July air temperature: validation by comparison with century-long meteorological records from northern Sweden. *Journal of Paleolimnology* 29(4): 475–493.

Larocque-Tobler, I., Heiri, O. & Wehrli, M. (2010) Late Glacial and Holocene temperature changes at Egelsee, Switzerland, reconstructed using subfossil chironomids. *Journal of Paleolimnology* 43(4): 649–666.

Larocque-Tobler, I. (2014) The Polish sub-fossil chironomids. *Palaeontologia Electronica* 17(1): 1–28.

Larocque, I., Pienitz, R. & Rolland, N. (2006) Factors influencing the distribution of chironomids in lakes distributed along a latitudinal gradient in northwestern Quebec, Canada. *Canadian Journal of Fisheries and Aquatic Sciences* 63(6): 1286–1297.

Laville, H. (1971) Recherches sur les chironomides (Diptera) lacustres du Massif de Néouvielle (Hautes-Pyrénées). *Annales de Limnologie-International Journal of Limnology* 7: 173–332.

Lencioni, V. & Rossaro, B. (2005) Microdistribution of chironomids (Diptera: Chironomidae) in Alpine streams: an autoecological perspective. *Hydrobiologia* 533(1–3): 61–76.

Lencioni, V. & Lazzara, M. (2004) I chironomidi subfossili (Diptera, Chironomidae) del Lago di Tovel (Trentino, Alpi centro-orientali): uno sguardo agli ultimi 400 anni. Studi Trentini di Scienze Naturali. *Acta Biologica* 81: 155–165.

Lencioni, V., Marziali, L., Rossaro, B. (2009) The first record of *Parochlus kiefferi* (Garrett, 1925) (Diptera, Chironomidae, Podonominae) from Italy. *Entomological News* 118(2): 127–133.

Lin, X.L., Stur, E. & Ekrem, T. (2018) Molecular phylogeny and temporal diversification of Tanytarsus van der Wulp (Diptera: Chironomidae) support generic synonymies, a new classification and centre of origin. *Systematic Entomology* 43(4): 659–677.

Lindegaard, C. (1992) Zoobenthos ecology of Thingvallavatn: vertical distribution, abundance, population dynamics and production. *Oikos* 64: 257–304.

Lotter, A.F., Heiri, O., Hofmann, W., van der Knaap, W., van Leeuwen, J., Walker, I. & Wick, L. (2006) Holocene timber-line dynamics at Bachalpsee, a lake at 2265 m a.s.l. in the northern Swiss Alps. *Vegetation history and archaeobotany* 15: 295–307.

Lotter, A.F., Hofmann, W., Kamenik, C., Lami, A., Ohlendorf, C., Sturm, M., van der Knaap, W.O. & van Leeuwen, J. (2000) Sedimentological and biostratigraphical analyses of short sediment cores from Hagelseewli (2339 m asl) in the Swiss Alps. *Journal of Limnology* 59(1s): 53–64.

Lukniš, M. (1973) *Relief of the High Tatra Mountains and their foreland.* Veda, Bratislava, 375 pp. (in Slovak).

Luoto, T.P. & Nevalainen, L. (2009) Larval chaoborid mandibles in surface sediments of small shallow lakes in Finland: implications for palaeolimnology. In

Palaeolimnological Proxies as Tools of Environmental Reconstruction in Fresh Water. Springer, Dordrecht p. 185–195.

Luoto, T.P. & Nevalainen, L. (2013) Climate change impacts on zooplankton and benthic communities in Lake Unterer Giglachsee (Niedere Tauern Alps, Austria). *International Review of Hydrobiology* 98(2): 80–88.

Luoto, T.P., Helama, S. & Nevalainen, L. (2013) Stream flow intensity of the Saavanjoki River, eastern Finland, during the past 1500 years reflected by mayfly and caddisfly mandibles in adjacent lake sediments. *Journal of Hydrology* 476: 147–153.

Magny, M., Aalbersberg, G., Bégeot, C., Benoit-Ruffaldi, P., Bossuet, G., Disnar, J.R., Heiri, O., Laggoun-Défarge, F., Mazier, F., Millet, L. & Peyron, O. (2006) Environmental and climatic changes in the Jura mountains (eastern France) during the Lateglacial – Holocene transition: a multi-proxy record from Lake Lautrey. *Quaternary Science Reviews* 25(5–6): 414–445.

Magyari, E.K., Vincze, I., Tóth, M., Hubay, K., Veres, D., Haliuc, A., Molnár, M., Jakab, G., Braun, M., Schäbitz, F. & Máthé, I. (2022) Palaeoenvironmental Changes During the Last Glacial Period in the Ciomadul Hills. In Ciomadul (Csomád), The Youngest Volcano in the Carpathians: Volcanism, Palaeoenvironment, Human Impact. Cham: Springer International Publishing p. 145–158.

Maheľ, M., (1986) *Geologická stavba československých Karpát. Paleoalpínske jednotky 1.* Veda, Bratislava, Geology of the Czechoslovakian Carpathians. Paleo-alpine Units 1 503 pp.

Marko, F., Sigdel, A., Bielik, M., Bezák, V., Mojzeš, A., Madarás, J., Papčo, J., Siman, P., Acharya, S. & Fekete, K. 2020. A comparison of Cenozoic Neo-Alpine tectonic evolution of the Western Carpathian and Himalayan orogenic belts (Slovakia – Nepal). *Mineralia Slovaca* 52 (2): 63–82.

Martin, J., Blinov, A., Alieva, E. & Hirabayashi, K. (2007) A molecular phylogenetic investigation of the genera closely related to Chironomus Meigen (Diptera: Chironomidae). *Contributions to the Systematics and Ecology of Aquatic Diptera. A Tribute to Ole A. Sæther. The Caddis Press, Columbus, Ohio*, p. 193–203.

Masteller, E.C., Wagner, R. (1984) The impact of sewage effluent on the occurrence of Psychodidae (Diptera) in a stream. *Freshwater Invertebrate Biology* 3(2): 98–104.

Matěna, J. & Frouz, J. (2000) Distribution and ecology of Chironomus Meigen species in the Czech Republic (Diptera, Chironomidae). In: Hoffrichter, O. (ed.). *Late 20th Century Research on Chironomidae: An Anthology from the 13th International Symposium on Chironomidae. Shaker Verlag, Aachen*, p. 543–548.

Matoshko, A.V. (2011) Limits of the Pleistocene glaciations in the Ukraine: a closer look. In: Ehlers, J., Gibbard, P.L., Hughes, P.D. (eds) *Developments in Quaternary Science. Elsevier*, p. 405–418.

Mauch, E. (2017) *Aquatische Diptera-Larven in Mittel-, Nordwest- und Nordeuropa: Übersicht über die Formen und ihre Identifikation*. Erich Mauch Verlag, 404 pp.

Meinander, M. (1996) Megaloptera Sialidae, alder flies. In Nilsson, A.N. (ed.), *Aquatic Insects of North Europe. A Taxonomic Handbook*, vol. 1. Apollo Books, Stenstrup. p. 105–110.

Michailova, P., Kownacki, A., Woznicka, O., White, K., Dean, A. & Szarek-Gwiazda, E. (2014) Macropelopia nebulosa group (Diptera, Chironomidae, Tanypodinae) – karyotype and morphology of larvae and pupae. *Zootaxa* 3852: 083–100.

Mihăiescu, R., Pop, A.I., Mihăiescu, T., Muntean, E., Beldean, S., Muntean, N., Alhafezl, L. & Ozunu, A. 2012. Physico-chemical characteristics of the karst lake Ighiu (Romania). *Environmental Engineering and Management Journal* 11(3): 623–626.

Moller Pillot, H.K.M. (1984) De larven der Nederlandse Chironomidae (Diptera): Inleiding, Tanypodinae & Chironomini. *Nederlandse Faunistische Mededelingen* 1(A): 1–277.

Moller Pillot, H.K.M. (2013) *Chironomidae Larvae of the Netherlands and Adjacent Lowlands. Biology and Ecology of the aquatic Orhocladiinae*. KNNV Publishing, Zeist, 312 pp.

Moubayed-Breil, J. & Bitušík, P. (2019) Taxonomic notes on the genus *Chaetocladius* (*laminatus*-group) (Diptera: Chironomidae, Orthocladiinae) II. Descriptions of *C. bitusiki* sp. n. and *C. mantetensis* sp. n., two relict species inhabiting cold alpine springs and streams. *Biologia* 74 (11): 1489–150.

Mráz, P. & Ronikier, M. (2016) Biogeography of the Carpathians: evolutionary and spatial facets of biodiversity. *Biological Journal of the Linnean Society* 119 (3): 528–559.

Mykitchak, T. (2016) Cladocera and Copepoda crustaceans of the Svydivets massif (Ukrainian Carpathians). *Visnyk of the Lviv University. Series Biology* 72: 149–160. (In Ukrainian with English summary).

Mykitchak, T., Reshetylo, O., Kostyuk, A., Popelnitska, O., Danilik, I., Tsarenko, P., Borsukevitch, L., Mateleshko, O., Martinov, O., Lilitska, G., Kapustin, D., Goncharenko, V. & Kokish, A. (2014) *Ecosystems of lenthic water bodies of Chornohora massif (Ukrainian Carpathians)*. ZUKC, Lviv, 288 pp.

Necsoiu, M., Mîndrescu, M., Onaca, A., & Wigginton, S. (2016) Recent morphodynamics of alpine lakes in Southern Carpathian Mountains using high-resolution optical imagery. *Quaternary International* 415: 164–174.

Nicacio, G. & Juen, L. (2015) Chironomids as indicators in freshwater ecosystems: an assessment of the literature. *Insect Conservation and Diversity* 8(5): 393–403.

Novikmec, M., Veselská, M., Bitušík, P., Hamerlík, L., Matúšová, Z., Klementová, B.R. & Svitok, M. (2015) Checklist of benthic macroinvertebrates of high altitude ponds of the Tatra Mountains (Central Europe) with new records of two species for Slovakia. *Check List* 11(1): 1522.

Nowicki, M., 1873a. Beiträge zur Kenntnis der Dipterenfauna Galiziens. *Krakau, Jagiellonische Universitäts Buchdruckerei*: 1–35.

Nowicki, M., 1873b. Beiträge zur Insectenfauna Galiziens. *Krakau, Jagiellonische Universitäts Buchdruckerei*: 1–55.

Oliver, D.R. & Roussel, M.E. (1983) Redescription of *Brillia* Kieffer (Diptera: Chironomidae) with descriptions of Nearctic species. *The Canadian Entomologist* 115(3): 257–279.

Oliver, D.R., McClymont, D. & Roussel, M.E. (1978) *A key to some larvae of Chironomidae (Diptera) from the Mackenzie and Porcupine River watersheds*. Fisheriers and Marine Service Technical Report no. 791, 73 pp.

Pecskay, Z., Kaličiak, M., Konečný, V., Lexa, J. & Žec, B. (2002) Geochronology of the Neogene volcanism in the Vihorlatské vrchy mountain range, Eastern Slovakia. *Geologica Carpathica* 53: 203–205.

Pinder, L.C.V. (1986) The larvae of Chironominae (Diptera: Chironomidae) of the Holarctic region – Introduction. In: Wiederholm, T. (ed.), Chironomidae of the Holarctic region. Keys and diagnoses. Part 2. Pupae. *Entomologica Scandinavica*, Suppl. 28: 5–7.

Pinder, L.C.V. (1995) The habitats of chironomid larvae. In: Armitage, P.D., Pinder, L.C. & Cranston, P.S. (eds), *The Chironomidae: biology and ecology of non-biting midges*. Chapman & Hall, 572 pp.

Pinder, L.C.V. & Reiss, F. (1983) The larvae of Chironominae (Diptera: Chironomidae) of the Holarctic region – Keys and diagnoses. In: Wiederholm, T. (ed.), Chironomidae of the Holarctic region. Keys and diagnoses. Part 1. Larvae. *Entomologica Scandinavica*, Suppl. 19: 293–435.

Pipík, R., Hamerlík, L., Hajnalová, M., Starek, D., Milovská, S., Milovský, R., Šurka, J. & Bitušík, P. (2022) Lake Morské oko (Vihorlat, Slovakia) was created in the Late Bronze Age (in Slovak). In Čiamporová-Zaťovičová, Z. & Laššová, K. (eds), *XIX. konferencia Slovenskej limnologickej spoločnosti a České limnologické společnosti, Zborník abstraktov*, 20.–24. jún 2022, Bratislava-Devín, 118 pp.

Pop, A.I., Mihăiescu, R., Mihăiescu, T., Muntean, E., Tănăselia, C., Maloş, C., Oprea, M-G. & Ozunu, A. (2012) Study on Bâlea and Călţun Glacier Lakes from Făgăraş Mountains. *ProEnvironment* 5: 260–265.

Quinlan, R. & Smol, J.P. (2001) Chironomid-based inference models for estimating end-of-summer hypolimnetic oxygen from south-central Ontario shield lakes. *Freshwater Biology* 46(11): 1529–1551.

Quinlan, R., Smol, J.P. (2010) Use of subfossil *Chaoborus* mandibles in models for inferring past hypolimnetic oxygen. *Journal of Paleolimnology* 44(1): 43–50.

Raddum, G.G. & Sæther, O.A. (1981) Chironomid communities in Norwegian lakes with different degrees of acidification. *Internationale Vereinigung für theoretische und angewandte Limnologie: Verhandlungen* 21(1): 399–405.

Raposeiro, P.M., Rubio, M.J., González, A., Hernández, A., Sánchez-López, G., Vázquez-Loureiro, D., Rull, V., Bao, R., Costa, A.C., Gonçalves, V., Sáez, A. & Giralt, S. (2017) Impact of the historical introduction of exotic fishes on the chironomid community of Lake Azul (Azores Islands). *Palaeogeography, Palaeoclimatology, Palaeoecology* 466: 77–88.

Rees, A.B., Cwynar, L.C. & Cranston, P.S. (2008) Midges (Chironomidae, Ceratopogonidae, Chaoboridae) as a temperature proxy: a training set from Tasmania, Australia. *Journal of Paleolimnology* 40(4): 1159–1178.

Reiss, F. (1968) Verbreitung lakustrischer Chironomiden (Diptera) des Alpengebietes. *Annales Zoologici Fennici* 5: 119–125.

Resh, V.H. & Cardé, R.T. (2009) *Encyclopedia of Insects*, second edition. Elsevier. 1132 pp.

Rieradevall, M. & Brooks, S.J. (2001) An identification guide to subfossil Tanypodinae larvae (Insecta: Diptera: Chrironomidae) based on cephalic setation. *Journal of Paleolimnology* 25: 81–99.

Rosenberg, D.M. & Resh, V.H. (1993) *Freshwater biomonitoring and benthic macroinvertebrates*. Chapman & Hall, NY, London, 487 pp.

Ross, D.H. & Merritt, R.W. (1987) Factors affecting larval black fly distributions and population dynamics. In Kim, K.C. & R.W. Merritt (eds), *Black flies: ecology, population management, and annotated world list*. The Pennsylvania State University Press, University Park: 90–108.

Rossaro, B. (1991) Chironomids and water temperature. *Aquatic Insects* 13: 87–98.

Rossaro, B. & Lencioni, V. (2015) A key to larvae of Diamesa Meigen, 1835 (Diptera, Chironomidae), well known as adult males and pupae from Alps (Europe). *Journal of Entomological and Acarological Research* 47: 123–138.

Rossaro, B., Solimini, A., Lencioni, V., Marziali, L., Giacchini, R. & Parenti, P. (2007) The relationship between body size, pupal thoracic horn development and dissolved oxygen in Chironomini (Diptera: Chironomidae). *Fundamental and Applied Limnology/Archiv für Hydrobiologie* 169(4): 331–339.

Rozkošný, R. (1980) *Klíč vodních larev hmyzu*. Academia, Praha, 521 pp.

Rubín, J., Galvánek, J. & Vydra, V. 1987. *Jewels of nonliving nature of Slovakia*. Osveta, Martin, 269 pp. (in Slovak with English, German and Russian Summary).

Rumes, B., Eggermont, H., Verschuren, D. (2005) Representation of aquatic invertebrate communities in subfossil death assemblages sampled along a salinity gradient of western Uganda crater lakes. In Segers, H. & Martens, K. (eds), Aquatic Biodiversity II. *Hydrobiologia* 542: 297–314.

Ruse, L. (2010) Classification of nutrient impact on lakes using the chironomid pupal exuvial technique. *Ecological Indicators* 10(3): 594–601.

Rück, A., Walker, I.R. & Hebda, R. (1998) A palaeolimnological study of Tugulnuit Lake, British Columbia, Canada, with special emphasis on river influence as recorded by chironomids in the lake's sediment. *Journal of Paleolimnology* 19: 63–75.

Sæther, O.A. (1975) Nearctic and Palearctic Heterotrissocladius (Diptera, Chironomidae). *Bulletin of the Fisheries Reseanth Board of Canada* 193: 1–67.

Sæther, O.A. (1976) Revision of *Hydrobaenus, Trissocladius, Zalutschia, Paratrissocladius*, and some related genera (Diptera: Chironomidae). *Bulletin of the Fisheries Research Board of Canada* 195: 1–287.

Sæther, O.A. (1979) Chironomid communities as water quality indicators. *Holarctic Ecology* 2: 65–74.

Sæther, O.A. (2005) A new subgenus and new species of *Orthocladius* van der Wulp, with a phylogenetic evaluation of the validity of the subgenera of the genus (Diptera: Chironomidae). *Zootaxa* 974: 1–56.

Sæther, O.A., Ekrem, T. (2003) Biogeography of afrotropical Chironomidae (Diptera), with special reference to Gondwanaland. *Cimbebasia* 19: 175–191.

Sæther, O.A., Ashe, P. & Murray, D.A. (2000) Family Chironomidae. In: Papp, L. & Darvas, B., (eds), *Contribution to a Manual of Palaearctic Diptera (with special reference to flies of economic importance). Appendix.* Science Herald, Budapest, p. 113–334.

Salmela, J., Härmä, O., Taylor, D.J. (2021) *Chaoborus flavicans* Meigen (Diptera, Chaoboridae) is a complex of lake and pond dwelling species: a revision. *Zootaxa* 4927(2): 151–96.

Samartin, S., Heiri, O., Vescovi, E., Brooks, S.J. & Tinner, W. (2012) Lateglacial and early Holocene summer temperatures in the southern Swiss Alps reconstructed using fossil chironomids. *Journal of Quaternary Science* 27(3): 279–289.

Serra-Tosio, B. (1973) Ecologie et bogéographie des Diamesini d'Europe (Diptera, Chironomidae). *Travaux du Laboratoire d'Hydrobiologie et de Pisciculture de l'Université de Grenoble* 63: 5–175.

Serra-Tosio, B. (1976) Chironomides des Alpes: Le genre *Pseudodiamesa* (Diptera, Chironomidae). *Travaux scientifiques du Parc national de la Vanoise* 7: 117–138.

Schmid, P.E. (1993) A key to the larval Chironomidae and their instars from Austrian Danube Region streams and rivers. With particular reference to a numerical taxonomic approach. Part 1. Diamesinae, Prodiamesinae and Orthocladiinae. *Wasser und Abwasser*, Suppl. 3: 1–514.

Siciński J. (2007) Ochotkowate (Chironomidae). In: Bohdanowicz, W., Chudzicka, E., Filipiuk, I. & Skibińska, E. (eds), *Fauna Polski – charakterystyka i wykaz gatunków. Vol. II*. Muzeum i Instytut Zoologii PAN, Warszawa, p. 30–32.

Silva, F.L. & Ekrem, T. (2016) Phylogenetic relationships of nonbiting midges in the subfamily Tanypodinae (Diptera: Chironomidae) inferred from morphology. *Systematic Entomology* 41(1): 73–92.

Slobodníková, V., Hamerlík, L., Wojewódka-Przybył, M., Sochuliaková, L., Szarlowicz, K., Buczkó, K., Chamutiová, T., Sedlačková Přidalová, M., Bitušík, P. (2023) Tracking fish introduction in a mountain lake over the last 200 years using chironomids, diatoms, and cladoceran remains. *Water* 15(7): 1372.

Smol, J.P., Birks, H.J.B., Last, W.M. (eds) (2001) *Tracking environmental change using lake sediments: Volume 4: Zoological indicators*. Springer Netherlands, 217 pp.

Soldán, T. (1988) Distributional patterns, host specifity and density of an epoitic midge, *Epoicocladius flavens* (Diptera, Chironomidae) in Czechoslovakia. *Věst. Čs. Společ. Zool.* 52: 178–289.

Soponis, A. (1977) A revision of the Nearctic species of Orthocladius (Orthocladius) van der Wulp (Diptera: Chironomidae). *The Memoirs of the Entomological Society of Canada* 102: 1–187.

Spies, M. & Sæther, O.A. (2013) Chironomidae. Fauna Europaea. In: Beuk, P. & Pape, T. (eds), Fauna Europaea: Diptera, Nematocera. Fauna Europaea version 2.6. Internet database: http://www.faunaeur.org (Accessed on 2019-11-10).

Stenson, J.A. (1981) The role of predation in the evolution of morphology, behaviour and life history of two species of *Chaoborus*. *Oikos* 37: 323–327.

Stoklasa, J., Dobríková, D., Sochuliaková, L., Pipík, R. & Hamerlík, L. (2017) Identifying white spots on the roadmap of Late Pleistocene and Holocene palaeolimnology in Slovakia: Review and future directions. *Biologia* 72(11): 1229–1239.

Stuchlík, E., Bitušík, P., Hardekopf, D.W., Hořická, Z., Kahounová, M., Tátosová, J., Vondrák, D. & Dočkalová, K. (2017) Complexity in the biological recovery of Tatra Mountain lakes from acidification. *Water, Air, & Soil Pollution* 228(5): 184.

Stur, E. & Ekrem, T. (2006) A revision of West Palaearctic species of the *Micropsectra atrofasciata* species group (Diptera: Chironomidae). *Zoological Journal of the Linnean Society* 146(2): 165–225.

Štefková, E. & Šporka, F. (2001) Long-term ecological research of high mountain lakes in the High Tatras (Slovakia). *Ekologia* 20(2): 101–106.

Svitok, M., Kubovčík, V., Kopáček, J. & Bitušík, P. (2021) Temporal trends and spatial patterns of chironomid communities in alpine lakes recovering from acidification under accelerating climate change. *Freshwater Biology* 66(12): 2223–2239.

Sweetman, J.N. & Smol, J.P. (2006) Reconstructing fish populations using Chaoborus (Diptera: Chaoboridae) remains – a review. *Quaternary Science Reviews* 25 (15–16): 2013–2023.

Szabó, Z., Buczkó, K., Haliuc, A., Pál, I., Korponai, J.L., Begy, R.C., Veres, D., Luoto, T.P., Zsigmond, A.R. & Magyari, E.K. (2020) Ecosystem shift of a mountain lake under climate and human pressure: A move out from the safe operating space. *Science of the Total Environment* 743: 140584.

Szabó, Z., Buczkó, K., Korponai, J.L., Luoto, T., Begy, R.C., Haliuc, A., Veres, D., Hamerlík, L., Csorba, R., Zsigmond, A.R. & Darabos, G. (2024) Two chironomid-inferred mean July air temperature reconstructions in the South Carpathian Mountains over the last 2000 years. *The Holocene* 34(7): 987–1004.

Szadziewski, R., Krzywiński, J. & Giłka, W. (1997) Diptera Ceratopogonidae, biting midges. In Nilsson, A.N. (ed.), *Aquatic insects of North Europe. A taxonomic handbook. Vol. 2. Odonata, Diptera*. Apollo Books, Stenstrup, p. 243–263.

Tarrats, P., Cañedo-Argüelles, M., Rieradevall, M. & Prat, N. (2018) The influence of depth and macrophyte habitat on paleoecological studies using chironomids: Enol Lake (Spain) as a case study. *Journal of Paleolimnology* 60(1): 97–107.

Tatole, V. (2023) An updated checklist of chironomid species (Diptera, Chironomidae) from Romania. *Travaux du Muséum National d'Histoire Naturelle "Grigore Antipa"* 66(1): 17–105.

Tátosová, J., Veselý, J. & Stuchlík, E. (2006) Holocene subfossil chironomid stratigraphy (Diptera: Chironomidae) in the sediment of Plešné Lake (the Bohemian Forest, Czech Republic): Palaeoenvironmental implications. *Biologia* 61(20): S401–S411.

Terek, J. (1981) Hydrobiological conditions of the Great Vihorlat Lake. *Biologia* 36: 339–352. (In Slovak with English summary).

Terek, J. (2009) *Lakes of Vihorlat*. University in Prešov, 143 pp. (In Slovak with English summary).

Teslenko, V.A. & Zhiltzova, L.A. (2009) *Key to the stoneflies (Insecta, Plecoptera) of Russia and adjacent countries. Imagines and nymphs*. Russian Academy of Sciences, Far Eastern Branch, Dalnauka, Vladivostok, 381 pp.

Tolonen, K.T., Brodersen, K.P., Kleisborg, T.A., Holmgren, K., Dahlberg, M., Hamerlik, L., Hämäläinen, H. (2012) Phantom midge-based models for inferring past fish abundances. *Journal of Paleolimnology* 47(4): 531–47.

Tóth, M., Buczkó, K., Specziár, A., Heiri, O., Braun, M., Hubay, K., Czakó, D. & Magyari, E.K. (2018) Limnological changes in South Carpathian glacier-formed lakes (Retezat Mountains, Romania) during the Late Glacial and the Holocene: A synthesis. *Quaternary International* 477: 138–152.

Tóth, M., Heiri, O., Vincze, I., Braun, M., Szabó, Z., Magyari, E.K. (2022) Limnological changes and chironomid-inferred summer air temperature from the Late Pleniglacial to the Early Holocene in the East Carpathians. *Quaternary Research* 105: 151–165.

Tóth, M., Magyari, E.K., Brooks, S.J., Braun, M., Buczkó, K., Bálint, M., Heiri, O. (2012) A chironomid-based reconstruction of late glacial summer temperatures in the southern Carpathians (Romania). *Quaternary Research* 77(1): 122–131.

Turis, P. & Jasík, M. (eds) (2007) *The Low Tatra Mountain National Park – natural values, history and current state of the protection*. Správa Národného parku Nízke Tatry, Banská Bystrica, 116 pp. (In Slovak with English summary).

Ujvárosi, L., Kolcsár, L.P., Bálint, M., Mihali, C. (2010) Pediciidae larvae (Insecta, Diptera) in the Carpathian Basin: preliminary results and further perspectives, *Acta Biol. Debr. Oecol. Hung.* 21: 233–246.

Uutala, A.J. (1990) *Chaoborus* (Diptera: Chaoboridae) mandibles – paleolimnological indicators of the historical status of fish populations in acid-sensitive lakes. *Journal of Paleolimnology* 4(2): 139–51.

Vallenduuk, H.J. (2017) Chironomini larvae of western European lowlands (Diptera: Chironomidae). Keys with notes to the species. With a redescription of Glyptotendipes (Caulochironomus) nagorskayae and a first description of Glyptotendipes (Caulochironomus) kaluginae new species. *Lauterbornia* 82: 1–216.

Vallenduuk, H.J. & Moller Pillot, H.K.M. (2007) *Chironomidae larvae of the Netherlands and adjacent lowlands. General ecology and Tanypodinae.* KNNV Publishing, Zeist, 144 pp.

Vallenduuk, H.J. & Morozova, E. (2005) *Cryptochironomus.* An identification key to the larvae and pupal exuviae in Europe. *Lauterbornia* 55: 1–22.

Varga, Z. (2011) Extra-Mediterranean refugia, post-glacial vegetation history and area dynamics in Eastern Central Europe. In: Zachos, F.E., Habel, J.E. (eds), *Biodiversity hotspots distribution and protection of conservation priority areas. Part 2.* Springer Verlag, Berlin, p. 57–87.

van den Berg, M.S., Coops, H., Noordhuis, R., van Schie, J. & Simons, J. (1997) Macroinvertebrate communities in relation to submerged vegetation in two Chara-dominated lakes. *Hydrobiologia* 342: 143–150.

van de Bund, W.J., Davids, C. & Spaas, S.J. (1995) Seasonal dynamics and spatial distribution of chydorid cladocerans in relation to chironomid larvae in the sandy littoral zone of an oligo-mesotrophic lake. *Hydrobiologia* 299(2): 125–138.

Vincze, I., Orbán, I., Birks, H.H., Pál, I., Finsinger, W., Hubay, K., Marinova, E., Jakab, G., Braun, M., Biró, T., Tóth, M. (2017) Holocene treeline and timberline changes in the South Carpathians (Romania): Climatic and anthropogenic drivers on the southern slopes of the Retezat Mountains. *The Holocene* 27(11): 1613–1630.

Vološčuk, I. (ed.) (1994) *Tatranský národný park.* Gradus, Slovakia, 551 pp.

Vološčuk, I. (ed.) (1999) *National Parks and Biosphere Reserves in Carpathians. The Last Nature Paradises.* ACANAP, Tatranská Lomnica, 248 pp.

Vondrák, D., Schafstall, N.B., Chvojka, P., Chiverrell, R.C., Kuosmanen, N., Tátosová, J., Clear, J.L. (2019) Postglacial succession of caddisfly (Trichoptera) assemblages in a central European montane lake. *Biologia* 74(10): 1325–1338.

Walker, I.R. (2001) Midges: Chironomidae and related Diptera. In: Smol, J.P., Birks, H.J.B. & Last, W.M. (eds), *Tracking environmental change using late sediments.* Vol. 4. Zoological Indicators, Kluwer Academic Publishers, Dordrecht, p. 43–66.

Walker, I.R., Smol, J.P., Engstrom, D.R. & Birks, H.J.B. (1991) An assessment of Chironomidae as quantitative indicators of past climate change. *Canadian Journal of Fisheries and Aquatic Sciences* 48: 975–987.

Waringer, J. & Graf, W. (2011) *Atlas der mitteleuropäischen Köcherfliegenlarven – Atlas of Central European Trichoptera Larvae.* Erik Mauch Verlag, 468 pp.

Warwick, W.F. (1991) Indexing deformities in ligulae and antennae of Procladius larvae (Diptera: Chironomidae): Application to contaminated-stressed environments. *Canadian Journal of Fisheries and Aquatic Sciences* 48: 1151–1166.

Wiederholm, T. (1980) Use of benthos in lake monitoring. *Journal of Water Pollution Control Federation* 52(3): 537–547.

Wiederholm, T. & Eriksson, L. (1979) Subfossil chironomids as evidence of eutrophication in Ekoln Bay, Central Sweden. *Hydrobiologia* 62(3): 195–208.

Wiggins, G.B. (2004) *Caddisflies, the Underwater Architects*. University of Toronto Press, Toronto, Buffalo, London, 292 pp.

Williams, N.E. (1988) The use of caddisflies (Trichoptera) in palaeoecology. *Palaeogeography, Palaeoclimatology, Palaeoecology* 62: 493–500.

Williams, N.E. (1989) Factors affecting the interpretation of caddisfly assemblages from Quaternary sediments. *Journal of Paleolimnology* 1: 241–248.

Wilson, R.S. & Ruse, L.P. (2005) A guide to the identification of genera of chironomid pupal exuviae occurring in Britain and Northern Ireland (including common genera from Northern Europe) and their use in monitoring lotic and lentic fresh waters. Freshwater Biological Association, CD-Rom.

Index

www.ingramcontent.com/pod-product-compliance
Lightning Source LLC
Chambersburg PA
CBHW071323230926
51849CB00002B/296
9789004694668